GENIE

GENIE

A SCIENTIFIC TRAGEDY

RUSS RYMER

HarperPerennial
A Division of HarperCollinsPublishers

First HarperPerennial edition published 1994.

Designed by Jessica Shatan

The Library of Congress has catalogued the hardcover edition as follows:

Rymer, Russ.
  Genie: a scientific tragedy / Russ Rymer —1st ed.
    p.   cm.
  ISBN 0-06-016910-9
  1. Genie, 1957– . 2. Abused children—California—Biography. I. Title.
RJ507.A29R95 1993
362.7′.6′092—dc20
[B]   92-53327

ISBN 0-06-092465-9 (pbk.)

  95 96 97 98 RRD/H 10 9 8 7 6 5 4 3

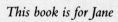

This book is for Jane

ACKNOWLEDGMENTS

I would like to thank some of those who nursed this book through its perilous birth, and the author through a protracted labor: My parents, Richard and Elizabeth Rymer, for their constant support and steel nerves. Melanie Jackson and Martha Kaplan, for their faith beyond reason that a manuscript would ever arrive, and Sara Lippincott, my editor at *The New Yorker,* for her sure and brilliant hand. Jane Palecek for shaping so many of these pages, and for enduring each of them, endlessly. Stephen Hall and Ed Dobb for their critical eye and sharp pencils. And Susan Faludi, for living so long in close expectance as this thing closed at last.

This book is owed especially to these three: Bill Sempreora, of Passaic, who was known as Bill Jacobs in the time he was my teacher, and Bill Cutler, of Atlanta, who was born Duncan Aswell, and who was my first editor. Both died in 1988, when they were still far too young for that, and both are missed. I wish they could have known what good fortune their counsel and example would bring me. And Bill Broder, my friend, who was the first to see this as a book, and who has since labored artfully that its author might make it a good one.

The borders separating reason and madness, the good and evil of our moral existence, are very haphazardly placed, here for the doctor, there for the moralist, and subject to change.

—ROBESPIERRE

Does one's integrity ever lie in what he is not able to do? I think that usually it does.

—FLANNERY O'CONNOR

CONTENTS

CONTENTS

I

FOUND

1

Sometime in the late seventh century B.C., it occurred to Psamtik I, the first of the Saitic kings of Egypt, to wonder which might be the original language of the world. Psamtik was, by all accounts, a forward-looking ruler. He was the first to open his country to large-scale immigration, receiving thereby a substantial infusion of Hellenic culture and, not incidentally, the Hellenic mercenaries with which he secured his reign against the claims of eleven rivals and against the Scythian, Ethiopian, and Assyrian armies on his frontiers. Considering that he undertook his scholarship between perennial military campaigns, it is not surprising that his interest in the language question had territorial overtones: the country possessed of the *lingua mundi* would own an indisputable hegemonic legitimacy. Yet he pursued his question with an unbiased rigor and a devotion to the scientific method which could be seen as admirably unsentimental, if not downright brutal.

As recounted by Herodotus from the vantage of two hundred years later, Psamtik's experiment was a simple one: two infants were taken from their mothers at birth and placed in the isolation of a shepherd's hut. The shepherd was instructed not to speak to them. They were reared thus on a diet of goat's milk and silence until one day two years later when, the shepherd returning to his hut, the pair accosted him with their first utterance. The word they had developed was "*bekos*," which, after some semantic inquiry on the part of the King, was determined to mean "bread" in the language of the Phrygians, an Indo-European people of Asia Minor. With the shepherd's account in front of him, Psamtik abandoned his nationalistic hopes and stood by the results of his research. He announced that Phrygian was the protolanguage and established himself as the protolinguist, the earliest practitioner of an enduring scientific pursuit.

Sadly—or perhaps fortunately, since except for the word *bekos* and a few texts and inscriptions little remains to us of the Phrygian language—Psamtik's research has not well stood the test of time. He has been accused of a certain methodological informality. There is no way of ascertaining whether, for instance, the children had a natural grasp of many languages and were merely expressing an innate human preference for Phrygian baked goods. Historians are satisfied that Phrygia was the birthplace of the flute and the Dionysian orgy, but probably not of human speech, and Psamtik is remembered by science mainly for his errors.

Nevertheless, in nearly every college primer on linguistics and in innumerable late-night conversations among practicing linguists, he is remembered. One such text, Vivien Tartter's respected 1986 *Language Processes,* has a two-sentence "Conclusion" that reads, "We still have a long way to go to understand language and its processing, and many exciting years of research ahead. But we have come a long way since Psammetichos!" The King's inclusion in the book, of course, like his more general durability, is evidence to the contrary: Psamtik is very much with us. While his experiment was flawed in fulfilling its declared intention, it was in other ways brilliant—an incisive bit of scientific prescience. It embodied both the theoretical questions and the practical quandaries that still bedevil the discipline. Beyond the arid statistics and the arcane analysis that characterize modern linguistics looms a philosophical question: What makes us special as a species? What part of our essential humanity is expressed in our ability to communicate with language? It is in that light that Psamtik's scientific sin— his experimentation on children—takes on the import that continues to so subtly trouble the science. For his sin was of the essence: in investigating one piece of the human charter, Psamtik, by his lack of compassion, did violence to another.

The science initiated by the Egyptian king has been revised and reinvented many times over the millennia, most recently in a Horn & Hardart on Woodland Avenue in Philadelphia, where Noam Chomsky began working out a set of ideas so revolutionary that their publication, in 1957, is referred to by some linguists as the Event. To its credit as a human endeavor, the science of linguistics has maintained through its generations a certain wistful indecision about its ambitions. It is a stalwart linguist—or an especially myopic one—who can avoid the temptation to look up from the voluminous tabulations of syntax and phonemics for an occasional glance into the heart of human nature, much the way astronomers look through the silica lens at the origins of time and the creation of the universe, the way physicists find God in the numbers.

Linguistics and astronomy constitute an unlikely sisterhood, for they are both often constrained to be more observational than experimental—astronomy because its subjects are too distant to be experimented on, linguistics because its subjects are too human. No longer are children impressed rudely from the crib, à la Psamtik, to serve as guinea pigs. But the revelations about how we acquire language still come from children—wild children, who have grown up with beasts as their only companions; abused or neglected children whose family histories replicate the isolation in the shepherd's hut, sometimes with far more attendant horror. The cases are exceedingly rare and generally fleeting. They become the property of whichever researcher is fortunate enough to be present at whichever dark hour.

In that regard, no subject has ever fallen into the lap of science out of more incomprehensible a world than the little girl who limped through the doors of a Los Angeles County welfare office in the fall of 1970, accompanied by her nearly blind and almost equally traumatized mother.

2

Temple City, California, is in many ways a typical town of the San Gabriel Valley, and Golden West Avenue, which runs due north through it, is a typical Valley residential street. In its straightness—for it is as straight as a surveyor's rod—you might divine a purpose, might suppose that its intended destination is the San Gabriel Mountains, whose shadowed canyons and snow-paneled peaks rise above the endless grid of suburban valley streets like the promise of a less confining world. Golden West Avenue never reaches the San Gabriels, near as they are.

It never escapes far into the more prosperous reaches of Arcadia. It is interrupted in its northward progress by other straight streets, wider and faster ones, and the San Gabriels remain a taunting vision, as distant in their way as the affluent hills of Hollywood, fifteen miles to the west.

Heading up Golden West from Las Tunas Drive, Temple City's main drag, you pass the parklike acreage of the city hall and citizens' center and the steep-roofed Christian Church. Then the public places are behind you and an orderly regime takes over of small homes, wooden or stucco for the most part, becoming more modest, shopworn, sunstruck, and insular block by block as you head north. Each house has a drive and a yard, and the yards are distinguished one from the other by a low, bright, chain-link fence or masonry wall or, more often, by a change in the texture of the grass.

Toward the Arcadia town line, a quintet of royal palms floats over the avenue like an incongruous apparition, their vapor trail trunks rising like rockets from curbside, their foliate starbursts a hundred feet above the ground. They are the street's only aristocratic flourish. For here there are no rolling estates, no guarded gates, no Armed Response medallions such as dot the curbs of Bel Air and Mulholland Drive. The equation of prominence and privacy that prevails in the wealthy precincts of Los Angeles is here turned on its head: security lies in a respectful anonymity—an injunction, in a land of compact privacies, to mind one's own business. People don't come to Temple City to be discovered, they come to be left alone. Golden West Avenue is above all a quiet street of quiet families. Before the disruption of that quiet in November 1970, the residents of one small house behind the row of royal palms were known to their neighbors as the quietest family of all.

The disruption was spectacular—enough so to earn a week's worth of stories in the *Los Angeles Times,* sandwiched between accounts of the trial of Charles Manson, the policies of Gov. Ronald Reagan, the acquittal of the My Lai massacre soldiers, and the bombing of Hanoi. "GIRL, 13, PRISONER SINCE INFANCY, DEPUTIES CHARGE; PARENTS JAILED," read the headline on November 17. The following day, a story headed "MYSTERY SHROUDS HOME OF ALLEGED CHILD PRISONER" featured a photograph of two men standing in a driveway: the girl's elderly, bespectacled father, clothed in rumpled khakis and a rumpled hat, one hand in his pocket and the other loosely holding a cigarette; and her brother, a tall teenager dressed in black, his arms folded and his face wadded in belligerent distress.

But it was another photograph that inflamed the public imagination and brought the curious cruising along Golden West Avenue in a slow, neck-craning procession that lasted the better part of a week. The photograph is of a girl's face— smooth, olive-shaped, pretty. A strand of dark hair has escaped from behind her ear to hang across her forehead. Her head is turned with an attentive tilt toward the camera, but her eyes do not meet the lens. She looks above us, as though some object of interest were hovering over the photographer's shoulder. Her expression gives nothing away. It is composed but not self-conscious, withdrawn but with no trace of sullenness. Her mouth, its full lower lip closed against the serrated curve of the upper in a perfect Cupid's bow, turns up at the ends in what might be the beginning of a smile, except that she is otherwise so serious, so pensive and watchful. The energy in her face is all in her eyes—without beseeching, they attract. If her face has an adult's earnestness, her eyes have the straightforward curiosity of a toddler, unburdened by any evident capacity for prejudice or appraisal. Her innocence is incongruous with the newspaper's report of the epic abuse she had suffered.

That her condition was cause for concern had been immediately apparent to the social worker who received her and her mother in the welfare office one morning in early November. Like much else in the child's history, her arrival there seems to have been a fluke. The mother had come seeking help not for the child but for herself; three weeks earlier, she had finally managed to flee an abusive marriage and was living nearby with her parents, who were all but destitute. Cataracts and a detached retina had rendered her 90 percent blind in her left eye and totally blind in the right. She was searching for the office for services for the blind but, leading her daughter by one hand and her aged mother by the other, she had stumbled mistakenly into the general social services office. The eligibility worker whom she approached was transfixed by the child, a small, withered girl with a halting gait and a curious posture—unnaturally stooped, hands held up as though resting on an invisible rail. The worker alerted her supervisor to what she thought was an unreported case of autism in a child she estimated to be six or seven years old.

The supervisor did not confirm the autism diagnosis but agreed that something was amiss. The ensuing inquiries found the girl to be a teenager, though she weighed only fifty-nine pounds and was only fifty-four inches tall. She was in much worse physical shape than at first suspected: she was incontinent, could not chew solid food and could hardly swallow, could not focus her eyes beyond twelve feet, and, according to some accounts, could not cry. She salivated constantly, spat indiscriminately. She had a ring of hard callus around her buttocks, and she had two nearly complete sets of teeth. Her hair was thin. She could not hop, skip, climb, or do anything requiring the full extension of her limbs. She showed no perception of heat or cold.

Of most interest to the scientists who were to become her constant companions was that she could not talk. What the social worker had mistaken for an autistic child's abstention from verbal communication was in fact a complete inability. The girl's vocabulary comprised only a few words—probably fewer than twenty. She understood "red," "blue," "green," and "brown"; "Mother" and some other names; the verbs "walk" and "go"; and assorted nouns, among them "door," "jewelry box," and "bunny." Her productive vocabulary—those words she could utter—was even more limited. She seemed able to say only "Stopit," and "Nomore," and a couple of shorter negatives.

The social worker paid a visit to the child's home and convinced the mother that her daughter needed attention. She was admitted to Childrens Hospital of Los Angeles for treatment of extreme malnutrition. An explanation for the child's state was eventually pieced together, thanks to the efforts of the Temple City police in the days following her discovery and to the persistent elaborations of scientists and social workers over the next several years.

A doctoral dissertation on the child, written by Susan Curtiss, a graduate student at the University of California at Los Angeles and the linguist who was to spend the most time with

her, begins, "To understand this case history, one must understand [the] family background." And, indeed, every scientist
involved with the unfortunate child would be drawn again and
again through that background, much as the rubberneckers
had been drawn down Golden West Avenue—hoping to find
in the neighborhood, in the house, and now in the story of the
household, some answer.

3

Like most personal histories, the child's preceded her by years.
Her parents had migrated to the Los Angeles area from different parts of the country, out of similarly desperate circum-
stances. Clark, her father, had grown up in foster homes and
orphanages in his native Pacific Northwest; Irene, her mother,
was from Altus, Oklahoma. Though Irene's upbringing seems
rock stable compared with Clark's, it could hardly be called
idyllic. She was so often left to be cared for by family friends
that she felt she had two sets of parents. She called the friends
Mamaw and Dadaw, but in that she was merely following an
established convention, for Irene was not the first generation of
her family that Mamaw and Dadaw had helped raise. Years earlier, when Irene's father, a teenager then, had been permanently thrown out of the house by his own natural father,
Mamaw and Dadaw had taken him in as well.

Growing up, Irene felt closer to her father than to her
mother, who seemed to her stern and unloving. Irene remembers one incident from her early years when her mother forced

her to help pull the laundry through an old-fashioned wringer washing machine. The crank handle slipped and hit the child in the head, causing neurological damage that would eventually contribute to blindness in one eye. Another of her memories is of processions of men passing the house carrying buckets. She asked where the men were going, and her mother said, "They're going to town to the soup line. They aren't as lucky as we are. We have food." The dustbowl had reached Oklahoma.

Irene's family was lucky indeed—her father had a job as night foreman at the cottonseed mill. But as the drought lengthened, they, too, decided to leave Altus. They headed west, to southern California, where Irene's father found work in a filling station. Their fortunes in their new state would never rise far above meager. Like other real-life Joads, they had run out of continent before reaching the promised land, and the family's three children approached maturity with little prospect except the assurance of a restricted future. When Irene was in her early twenties, she found a traditional solution for her predicament and, traditionally, her parents opposed it. The man she married was twenty years her senior.

They met in Hollywood, in a drugstore where Irene was employed behind the soda fountain and where Clark would stop by on occasion to chat with the druggist about the horses. Clark was unemployed, but that didn't last long—the war was on and every hand was needed. The government impressed him into a profession for which he was suited by inclination, for despite his lack of schooling, Clark loved mathematics and devoured whatever books he could find on algebra, trigonometry, and geometry. The government, in Irene's words, "freshened him up" on his math, and turned him into a good machinist, putting him to work in the aircraft assembly lines in Santa Monica. In a photograph taken during their early years together, Irene and Clark appear to be a happy couple, even a bit glamorous. They are leaning against a shiny black sedan; Clark's crisp fedora is tipped onto the back of his head as he and his wife turn to each other with broad and mutual smiles.

After V-J Day, Clark parlayed his newfound skills into a good job with the aircraft industry and proved good at it. He bet moderately on the horses at the nearby Santa Anita race-track. He and his bride watched their money and enjoyed listening to radio shows. But surface felicities aside, Irene had run headlong out of a confining upbringing into a confining marriage. Clark was jealous of her least attention to others; she generously describes his attitude as "overly protective." She has said that her life came to an end on her wedding day.

Prominent among Clark's restrictions was his express desire not to have children. For one thing, they were noisy. Late in Irene's first pregnancy, five years into their marriage, Clark beat her severely. In the hospital for treatment of her injuries, Irene went into labor and gave birth to a healthy daughter. The infant's crying infuriated Clark, and she was placed in the garage, where, at the age of two and a half months, she died. Irene contends that the girl had been put there only to spare her the noise while the linoleum was being removed from the kitchen floor, and that once in the garage she had been struck with "quick pneumonia." The likelihood is that behind the euphemism was a case of death by exposure. A subsequent infant was more literally a victim of the couple's incompatibility: it died of Rh blood poisoning soon after birth. Irene's third pregnancy produced a healthy son. He survived infancy, but his development was stifled by an approximation of the neglect that had killed his oldest sibling. He was slow to walk and at three years of age was not yet toilet-trained. However, he was saved by the intercession of his paternal grandmother, who took him in and kept him for several months, long enough to get him back on track. In April 1957, Clark and Irene had their fourth child, a girl. She, too, had Rh blood poisoning, but she was given a transfusion soon after birth.

Back home with their new daughter, Clark and Irene received a package postmarked Oklahoma. It contained a Bible. Inside the cover was a note of congratulation, bearing the words, "Dear Irene, We was so glad to hear of your little girl. Now you have a pair. We wish we could send her something fine but I don't know of anything that will help her through life any better than this little book it will be a lamp to her foot steps. May God Bless you as her mother and may she be a fine girl for Jesus is my prayer." The gift was from Mamaw, who had watched over two generations of the family and was now wishing the best for a third. There were early indications that Mamaw's prayer would go unheard—the growing girl suffered the same developmental fate as her older brother, lagging behind in her habits and physical stature, and this time there was no paternal grandmother to rescue her at the critical moment.

Clark had an extraordinary attachment to his mother, surprising considering how little of his childhood had been spent with her. She was a flamboyant woman—at one time, she had managed a brothel—and was given to traveling armed. The prodigal mother seemed intent on making up for her inattention to Clark's upbringing by doting on him in his middle age. Until he picked up the machinist's trade, she had supported him. Even afterward, she helped pay his bills and frequently drove over to his and Irene's house to help out—in Irene's estimation, making a pest of herself. When they were together, Clark and his mother argued. For one thing, she thought her son intolerably straitlaced. Nevertheless, he was slavishly devoted to her, to the point where Irene never became more than a secondary allegiance in his life.

During one of her visits, in December 1958, Clark's mother was struck by a car and killed as she crossed the street with her grandson to buy an ice-cream cone. Irene called Clark at work. He rushed to the hospital, but his mother was almost beyond identification—in its frantic escape, the car had dragged her a long ways down the street. A teenager was arrested the next day and charged with hit-and-run and drunken driving—he received a probationary sentence. The court's leniency fueled Clark's fury.

After the accident, Irene recounts, things started changing. Clark's transformation was later described to me by Jay Shurley, a professor of psychiatry and behavioral science at the University of Oklahoma who was called in to the case and got to know what remained of the family, though he came too late to meet Clark. "Clark went beyond grief," Jay Shurley told me. "His depression began feeding on itself, on his isolation. The external world had given him a signal that he didn't count, his mother didn't count. Clark was very serious minded. He allowed himself no leeway to get around problems. He wasn't even religious in a way that would have helped him deal with trauma. He became enmeshed in his own withdrawal. His mistrust went beyond reality."

Clark decided that a world without his mother, a world that did not care to adequately punish her murderer, was a world he could best do without. He quit his job and moved his family into his mother's two-bedroom house on Golden West Avenue, where he would live out the last decade of his life as a recluse, with his family as virtual prisoners.

4

Irene's world closed in on her severely at this time. Her encroaching blindness made her almost completely dependent on her tormentor. Their son was allowed out of the house to attend school or to play with a neighbor but for little else, and within the house he was effectively a hostage. He slept on a pallet on the living room floor; his parents also slept in the living room—his mother on the floor and Clark in an easy chair in front of a defunct television set, sometimes with a gun in his lap. The main bedroom, according to some accounts, was kept undisturbed as a shrine to Clark's mother. But it was the daughter—twenty months old when the family moved—who bore the brunt of Clark's renunciation. "In essence, Clark appointed himself a guardian to his family," Jay Shurley told me. "His delusion was that his daughter was retarded and was going to be very vulnerable to exploitation. He dreaded the idea of people taking advantage of her."

After one of the child's rare early medical examinations, a pediatrician noted on her records that she was "slow," and pronounced her a "retarded little girl with kernicterus"—a condition that sometimes results from a botched transfusion for Rh incompatibility. "Clark amplified that to delusional intensity—that this girl was profoundly retarded," Shurley told me. "He was convinced that she would need his protection from the evil of the world, and that no one was better prepared than he to recognize its evil. He didn't reckon, of course, on his own evilness. These people never do."

Clark's idea of protective custody is described in Susan Curtiss's doctoral dissertation, which was published as a book—*Genie: A Psycholinguistic Study of a Modern-Day 'Wild Child'*—in 1977 by Academic Press. In both the dissertation and the book, the girl is referred to not by her real name but by her scientific alias, Genie—the name used in the symposium papers, the psychology magazines, and the textbooks and contrived in order to protect the child's identity. Curtiss's account agrees with that of other investigators. She wrote:

In the house Genie was confined to a small bedroom, harnessed to an infant's potty seat. Genie's father sewed the harness, himself; unclad except for the harness, Genie was left to sit on that chair. Unable to move anything except her fingers and hands, feet and toes, Genie was left to sit, tied-up, hour after hour, often into the night, day after day, month after month, year after year. At night, when Genie was not forgotten, she was removed from her harness only to be placed into another restraining garment—a sleeping bag which her father had fashioned to hold Genie's arms stationary (allegedly to prevent her from taking it off). In effect, it was a straitjacket. Therein constrained, Genie was put into an infant's crib with wire mesh sides and a wire mesh cover overhead. Caged by night, harnessed by day, Genie was left to somehow endure the hours and years of her life.

There was little for her to listen to; there was no TV or radio in the house. Genie's bedroom was in the back of the house next to [the master] bedroom and a bathroom. . . . The father had an intolerance for noise, so what little conversation there was between family members in the rest of the house was kept at a low volume. Except for moments of anger, when her father swore, Genie did not hear any language outside her door, and thus received practically no auditory stimulation of any kind, aside from bathroom noises. There were two windows in her room, and one of them was kept open several inches. She may, therefore, have occasionally heard an airplane overhead or some other traffic or environmental noises; but set in the back of the house, Genie would not have heard much noise from the street.

Hungry and forgotten, Genie would sometimes attempt to attract attention by making noise. Angered, her father would often beat her for doing so. In fact, there was a large piece of wood left in the corner of Genie's room which her father used solely to beat her whenever she made any sound. Genie learned to keep silent and to suppress all vocalization. . . .

Just as there was little to listen to, there was not much for Genie to touch or look at. The only pieces of furniture in her room were the crib and the potty seat. There was no carpet on the floor, no pictures on the walls. There were two windows, but they were covered up except for a few inches at the top out of which Genie could see the sky from one and the side of a neighboring house from the other. There was one dim, bare ceiling light bulb, a wall of closets, and another wall with the bedroom door. The room was a dirty salmon color. Occasionally, two plastic raincoats, one clear and one yellow, hung outside the closet in the room, and once in a while Genie was allowed to "play" with them. In addition, Genie was sometimes given "partly edited" copies of the TV log, with pictures that her father considered too suggestive removed (like women advertising swimming pools, etc.). She was also given an occasional empty cottage-cheese container, empty thread spools, and the like. These were Genie's toys; and together with the floor, her harness, and her body, they were her primary sources of visual and tactile stimulation.

Genie's diet was equally limited. She was given baby foods, cereals, an occasional soft-boiled egg. Under pressure from the father to keep contact with Genie to a minimum, she was fed hurriedly, usually by having food stuffed into her mouth. Should Genie choke and spit out some of her food, she would have her face rubbed in it. . . .

Genie's father was convinced that Genie would die. He was positive that she would not live past the age of twelve. He was so convinced of this that he promised his wife that if the child did live beyond twelve, the mother could seek help for Genie. But age twelve came and went; Genie survived, but the father reneged on his promise. The mother, too blind to even dial the phone and forbidden under threat of death to contact her own parents (who lived in the area), felt helpless to do anything.

Finally, when Genie was 13½ years old, Genie's mother, after a violent argument with her husband in which she threatened to leave unless he called her parents, succeeded in getting her husband to telephone her mother. Later that day Genie's mother took Genie and left her home and her husband.

Curtiss went on to relate the discovery of the girl: how she was taken into custody by the police; how the parents were arrested and charged with child abuse; how the child was admitted to the hospital. The family history is wrapped up, like Little Dorrit's, with a breath of exultation: "She had been discovered, at last."

But the real epitaph to the era was written by Clark himself. On the morning of November 20, 1970—the morning that he and his wife were to appear in court on charges of willful abuse or injury to the person or health of a minor—he spread out a blanket and a sheet of cellophane on the living room floor and shot himself through the right temple with a .38 caliber Smith and Wesson revolver that had once belonged to his mother. He was seventy years old. He left two notes, scrawled with a ballpoint pen. One was for the police and it read, in part, "My son is out in front with friends. He hasn't the slightest idea of what is going to happen." The second was to his son and included these instructions:

Don't take that shirt back. It's for my funeral. You know where my blue shirt is. Underwear in hall closet. I love you. Goodbye and be good.

—Dad

Clark did not leave a note for his wife or his daughter, but he did include in his farewells a sentence that seemed addressed to the public at large: to the press that had exposed his family's disarray; to the people in the automobiles, whose finger-pointing promenade had distressed him tremendously; to the scientists and doctors who had taken his daughter and renamed her. The sentence is as much a curse on the scientists' future efforts as it is an oblique defense of Clark's own past actions. He wrote, "The world will never understand."

Already in court that morning, Irene had heard her counsel enter a plea of not guilty, on the ground that she had been forced into her role by an abusive husband. The judge received a message and summoned the lawyers into chambers. Irene's counsel returned to tell her that her husband was dead. She was visibly shaken, the lawyer later recalled, but did not break down. "She just sat there, silent," he said. In a subsequent session, her plea was accepted.

Clark's suicide—reported, like the parents' arrest, on network news—did nothing to lessen interest in the case. The press had set up camp around Childrens Hospital, where Genie was now residing. Childrens was, and is, one of the most prominent and up-to-date pediatric facilities on the West Coast, and one well versed in the particular security concerns Genie presented, for, though its surrounding neighborhood and many of its patients are poor, it has also had among its clientele the children of Hollywood celebrities. Freed from her little room and placed in the most competent of professional hands, Genie was, in the view of the doctors and psychologists and others who were now becoming involved with her progress, liberated. If such a thing was possible, she was to be given a chance at a new life, with new surroundings, a new future—even a new mission—to go along with her new name.

5

Room 2113 of Campbell Hall on the campus of the University of California at Los Angeles—the office of Susan Curtiss in the summer of 1988, when we first met—is the type of airless, overinhabited cul-de-sac that, were it a street in Paris, would be labeled in the guidebooks as an impasse. In academe, it is labeled something else: if not quite a boulevard, at least a respectable avenue.

Curtiss had risen through a succession of such offices to become an associate professor of linguistics at UCLA. She was sharing her small space with two of her graduate students. Her desk was crammed into a far corner of the room, and over it were several pictures tacked to an orange room divider. There were photographs of her two daughters, aged five and one, and there was a drawing of Curtiss herself, done by Genie almost fifteen years earlier. The drawing was a stick figure, made with a series of quick crayon strokes. It wasn't easy to decide whether the rendering was immature for an artist in her middle teens or, in a primitivist way, accomplished, for its portrayal of its subject was accurate: Curtiss is tall, twig skinny, and as nervous as summer lightning. She is also extraordinarily focused, in the ironclad manner of one who has long done battle with the hectoring distractions of the academic world.

In 1971, when Genie entered her life, Curtiss was twenty-six years old and a first-year graduate student in the linguistics department. "I was one of the few linguists on campus studying language acquisition in children," she told me. "It seemed to me that once we came to understand language acquisition, we would have answers to most of the central questions of linguistics. Besides, I love children. It seemed as if it would be fun to have them be my source of data."

Her interests had put her in the right place at the right time, to say the least. She remembers the spring afternoon when she was summoned into the office of her faculty adviser, Victoria Fromkin. Fromkin, who became dean of the linguistics department and is now a professor emeritus, began discussing developments in a case of an abused and linguistically deprived child.

Curtiss had already heard of the case. "There had been a lot of press," she remembered. But now she was being invited in, on the ground level. "This was before the first research grant," she told me. "Before a lot of facts about the case were learned. As a new student, I found myself presented with an opportunity that changed my life in every way, personally as well as academically. Because the case is an important one, it shaped my future research, right down to today. I was just starting on the core curriculum then. I hadn't been exposed to many of the issues that Genie presented to me. I wasn't even aware of the critical-period hypothesis."

In 1971, the science of linguistics was perplexing even to some of its old hands. The critical-period hypothesis—the idea that there are certain distinct periods in a person's development during which skills like a first language can be learned—was just one of a host of new contentions. As the questions changed rapidly, there was also a shift in who was asking them. Curtiss's field, the acquisition of language by children, had previously been the carefully guarded purview of psychology departments—that is, for the duration of the twentieth century but not much earlier than that. Arguably, linguistics is to science what the Monongahela River is to North America: the most contested property in the realm. It is soaked with the blood of poets, theologians, philosophers, philologists, psychologists, biologists, and neurologists, along with whatever blood can be got out of grammarians. Each discipline has at one time or another set its flag in the territory, knowing that its internal orthodoxies would be partly determined by whoever owned the language question. Susan Curtiss was in the vanguard of the newest of a hundred raiding parties.

Until the High Renaissance, European philosophers had related the language question, along with most others, to the Bible. Any human attribute, it was thought, must be as inevitably mysterious and divine beyond investigation as the creator it reflected. Then Descartes committed the heresy of hacking man in half: he proposed the independence of the soul from the body, of the mind from the brain. This dualism allowed much leeway to the nascent science of biology. If the mind and soul were yet sacrosanct, one could at least put the scalpel to their corporeal counterparts. There was impressive historical testimony in favor of including the study of language in this new, naturalist science of biology. In the third century B.C., Epicurus, the first Greek philosopher to address the origins of language, felt that it was the creation not of God or of human intellect but of a far less interested party: nature. Language, he said, was a biological function, like vision or digestion. But his view was anathema to the tenor of later times, when language was considered an integral part—perhaps the keystone—of man's soul or (less likely) man's reason, or both. In the late seventeenth century, Leibniz proclaimed language ability to be a gift of God, with its form of expression determined by natural instinct—except for Chinese, which, he suggested, was the invention of a wise man. Thus linguistics was left standing with one foot on the theological dock and the other in the naturalist boat.

The discomfort was relieved somewhat by the rise of the social sciences at the end of the eighteenth century. If language was somewhere between theology and biology, then perchance it could be considered a problem for anthropologists, with linguists playing a backup role. The voyages of exploration and colonization had shaped the public imagination the way the Crusades shaped that of earlier times, but with a more utilitarian grail. Richard the Lion-Hearted had been vanquished by Sir Richard Burton. Linguists quit worrying themselves with questions of the Vulgate text and got busy cataloguing newfound languages, by the hundreds. (Their efforts continue today in university labs and rain forests around the world. Some of the most adroit work for Bible societies. They are sent into uncharted regions to translate the Bible for newly discovered tribes. In a few hours' time they can be conversational in a new language, in a day or two they can be fluent, and after several weeks they can emerge from the jungle and begin at the Beginning in Hua or Yanomamö.)

By the late nineteenth century, as the comparative linguists handled the relationship of one tongue to another, the bulk of the questions concerning the relationship of language to man had disappeared into psychology—a discipline the questions helped create. And that's where they more or less stayed until "the Event"—the publication of Noam Chomsky's *Syntactic Structures* in 1957, the year of Genie's birth.

6

The galvanic effect of Chomsky's innovation was described to me once by Catherine Snow, a professor of human development and psychology at Harvard University. "There was a barrenness in the study of language acquisition through the 1940s and fifties," she said. "In the last issue of the *Handbook on Developmental Psychology* before Chomsky's breakthrough there was a chapter by a very competent and very respected woman whose work is all about vocabulary. Until 1957, linguists believed that was most of what there was to think about: vocabulary. Then Chomsky made syntax central, and for the first time it was clear how big the task was, how difficult the child's task was in acquiring language. To have this new conception of what language was made the questions all the more compelling, interesting. It was like driving across the prairie and all of a sudden you see the Rocky Mountains jump out at you."

Chomsky's syntax bore little resemblance to the rules of grammar so ruefully learned by seventh graders. Those rules lie on the surface of language—Chomsky wanted to know what hid within. When he and his compatriots began isolating and defining some inner rules and using them as a lens through which to view language, they found that the great variety of surface structures could be distilled into a set of core principles. I had heard part of this distillation, on the sentence level, explained by Lila Gleitman, head of the linguistics department at the University of Pennsylvania.

"Oversimplifying," she had told me, "we can say that Chomsky said there is one sentence we are able to understand, but most of the sentences we hear are not like that model. So we have to move the sentence around to understand it. This is called 'movement' by linguists. The sentence is a mystery. It has a perpetrator and a victim, and often a blunt object, which is usually 'the.' The task of the language learner is to solve the mystery."

Movement not only condensed the infinite variety of sentences within a language, it also implied a radical thought: that even the sentences of diverse languages—of Japanese, with its inverted phrases; of Finnish, which expresses cases the way Latin does; of Lithuanian, among modern European languages the one closest to Sanskrit; of Spanish, in which the subject of a sentence is commonly omitted—are not fundamentally different from English sentences.

Some linguists have speculated, basing their hypothesis chiefly on similarities of vocabulary and pronunciation, that all languages derived from a common ancestor. Chomsky doesn't think so. On the syntactic level that he is concerned with, languages don't just have similarities—they are identical. The source of such an astounding uniformity, Chomsky argues, must be sought closer to home than an ancient protolanguage.

It must be contained within us—within the species. The inner rules of language—what Chomsky calls "universal grammar"—are either the product of an unparalleled achievement of human cognition or ingrained on a level more basic than thought. Chomsky opted for the latter explanation: we don't learn the inner rules of language, he said, we're born with them. The question was no longer simply "How is language designed?" but "How does language reflect the way *we* are designed?"

Chomsky's labors have earned detractors as well as adherents. Every modern linguist carries, involuntarily and sometimes unfairly, a vest-pocket vita summarizing his life's work as pro-Chomskian or anti-Chomskian. There are those who object to Chomsky because of his prominence in the field, and those who object to his prominence outside of it, in endeavors such as politics and philosophy and engineering. But most of the contention centers on theory. The school of linguistics associated with Chomsky's ideas—a school described variously as "nativist," "generative," "transformational," "innatist," and "rationalist"—quickly met with heated opposition from the school of "environmentalists," or "empiricists," who hold that a child learns language from its interaction with the world and from the speech of its parents.

Both schools have since fragmented and their ideas and observations have mingled over the years, and these days the contest, from without, looks decidedly intramural. As Gene Searchinger, a documentary filmmaker, once told me, "Any one of those people is operating on Chomskian precepts." Searchinger has spent the last ten years in his studio on New York's Seventy-second Street ("over the fruit stand and next to the OTB") making a series of films about linguists, a project so extensive that he has caused some concern that the next raid on the language question is about to be mounted by filmmakers. "I love the pro- and anti-Chomsky debate," Searchinger said. "It reminds me of the joke where the guy says, 'I don't like so-and-so, he's a communist.' And the other guy says, 'He's not a communist. He's an anticommunist.' And the first guy says, 'I don't care what kind of communist he is, I still don't like him.'"

Chomsky's detractors have mounted any number of academic assassination attempts, many of them imaginative and most begrudgingly affectionate. (One researcher found a way to get his coinage into the permanent scientific record, which rarely has a place for satire, by christening one of his primate subjects "Nimchimpsky.") It has been decidedly hard to work in the field during recent decades and not work in Chomsky's shadow.

Among those in the field who have led me through the perils of linguistic theory, Catherine Snow is an avowed "non-Chomskian" environmentalist, and Lila Gleitman a stalwart nativist. Gleitman refers to the time before Chomsky as "B.C." "He is a towering figure of twentieth-century science," she told me. Gleitman came to Penn as a student, soon after Chomsky had left, in an era still variously wedded to old ways: automobile bumpers were still painted black because chrome offended the Amish, and she was not allowed to pursue her dissertation into the experimental realm because a linguist doing experimentation might offend the university's psychologists. She studied in the same Horn & Hardart (since demolished for a library) where Chomsky had worked on his theories.

"I sat in there all of one summer practicing my Thucydides," she told me, "stumbling, stumbling, stumbling. I wanted to be a classicist, but I was beyond the critical period." Fortunately, classics led her to Herodotus, who taught her something about academic resurrection. "You know, when people were reading Herodotus, he recounted that the way the Egyptians planted seeds was to go down the row and have little pigs follow them, and the pigs would push the seeds in with their little hooves." She snorted with delight over this image and made little hooves out of her hands and pistoned them up and down. "People thought that was one of Herodotus's real howlers: there *were* no pigs in ancient Egypt. Then they opened up this one tomb and on one of the friezes people were sowing seeds and being followed by little pigs. He was right, after all."

Herodotus also led Gleitman to Psamtik, and thus to linguistics, and thus to Chomsky, whose mentor at the University of Pennsylvania became her faculty adviser. "Chomsky was Zelig Harris's best student," she said. "One of his best students. I was another, but it's blushable to include oneself in the same breath as Chomsky. The rest of us, we're victims of a modern Augustus. We're little pigs, following along behind," and she made the piston motion with her hands again, but this time didn't laugh, "pushing in the seeds."

Sometime during her first couple of years in the linguistics program, Gleitman asked her adviser about his celebrated pupil's ideas. Harris had fallen out with Chomsky, and the split had the seismic effect on the linguistic profession of John the Baptist falling out with he who came after. When Gleitman asked Harris about Chomsky, he told her, "The trouble with Noam is he thinks there is one language and one god, and I think that's either too few or too many."

7

Since the mid-1950s, Chomsky has taught at the Massachusetts Institute of Technology. His office is in Building 20, a shabby, three-story frame structure behind barbed wire on the edge of the campus. The building is a holdover from MIT's more provisional days, about the only example of its type that hasn't been replaced by something new and shiny. Its deceptive humility is appropriate to a deceptive past. During World War II it was used for the development of radar; the scientists working there then called themselves radiologists in order not to attract German job offers. But the "radiologists" have long since relinquished the warren of confused corridors, unplastered drywall, and uncovered pipes and cables to others, among them linguists. There are five wings, extending like fingers—from above, the building must look something like a baseball mitt drawn on an Etch-A-Sketch. At the end of one of the fingers are the cluttered couple of rooms that Chomsky calls home—the larger one for the secretary, the smaller one with a hook rug on the floor, a map of Italy pinned to the wall, and Chomsky's desk facing a window with a view of the faculty swimming pool.

I caught up with Noam Chomsky not in Building 20, but in room 250 of Building 10, a steeply pitched amphitheater classroom with an orchestra pit full of sliding blackboards. He was sitting in a pink chair in the front row, speaking into one of Gene Searchinger's movie cameras. "Recently, this rather common auditorium was filled with many young linguists debating the central issues of the science," he said. "Thirty years ago, the number of people who could even have conceived of these questions was virtually nil."

Searchinger yelled "Cut!" and the camera went dead. Chomsky, a shy matchstick of a man, crumpled back into his chair and began chatting with Searchinger while the crew adjusted the lights. Searchinger had the appearance of a stockbroker on two telephones.

GRIP (to Searchinger, yelling): "Is that good?"

SEARCHINGER: "Yes. No. Move it up."

CHOMSKY (to Searchinger): "What's more sacrilegious than religion?" (Grip raises lights.)

CAMERAMAN (to Searchinger): "The chair back is lit. Is that what you want?"

SEARCHINGER: "That's OK."

CHOMSKY (to Searchinger): ". . . but perfection? There's no such thing, unless you're religious."

CAMERAMAN (to Searchinger): "He's got a halo. Is that OK?"

SEARCHINGER: "That's OK, too."

Finally, Searchinger said "Sticks," a slate marked TAKE 5 was held in front of Chomsky's face and snapped shut, and Chomsky returned to the subject of his life's work and Searchinger's film.

"Language is a tool," he said. "The tool is endlessly useful—

in the sense that we commonly create and understand sentences that we have never heard before. How do we do it? Language is like a hammer: it can be used in many ways, and what it does depends on the person using it. Nevertheless, it is a system with a structure. Anything with structure has to have limits. It must; otherwise, it wouldn't work. If a hammer were an amorphous blob, it would not be useful.

"The problem arises when you look carefully at that structure—when you start to take language seriously. If you have succeeded in finding some structure, you've just begun. You're ready to ask new questions of the world. There was a basic assumption of the study of language and human behavior in the 1950s—that we should concentrate on what people do and produce. There is a major new perspective: a shift in focus to the inner mechanisms of mind that account for behaviors. What are the inner mechanisms?"

Among the telling mysteries the innatists found when they looked carefully at those inner mechanisms was one sometimes called the Scandal of Induction. Simply put: How are children able to develop a rule system not only for the finite number of sentences they've heard, but for all the infinite sentences they may be called upon to make? In fact, children early on respond to rules so arcane that adults rarely invoke them and could not, if they had to, explain them.

I had heard some specific examples of these underlying rules from avowed Chomskian and fellow MIT linguist Steve Pinker. Young children, he told me, understand that verbs are divided into classes. One aspect that segregates them is whether their action happens directly (as in "pushing") or ballistically (as in "throwing"). Children know that you can reverse a sentence in which the action is ballistic, so that "Kick the ball to John" becomes "Kick John the ball." But you cannot reverse "Push the ball to John" to "Push John the ball" because guided, hands-on motions follow a different syntactic rule. Likewise, you can reverse the sentence "I gave him a book" to "I gave the book to him." But you cannot change "I donated the painting to the museum" to "I donated the museum a painting." Why? Because Anglo-Saxon verbs like "give" are of a different class than the Latinate verbs that William the Conqueror brought to England from the continent in 1066. You can "throw John the ball" but you can't "propel him the ball."

Remarkably, children recognize the class distinction between the generally one-syllable native words and the longer, Latinate ones. They know you can shake something up but not vibrate it up. In tests, children three and a half years old even treat nonsense verbs like "pilk" differently than nonsense verbs like "ogulate."

Still within the realm of verb oddities is one that might be called the ballpark anomaly. There is a whole class of verbs that take a different past tense than others that sound the same. For example, the past tense of "fly" is "flew," yet, as Pinker pointed out to me, the baseball player didn't flew out, he flied out. The goalie wasn't high-stuck, he was high-sticked. Americans in particular like to make verbs out of nouns like "stick" and "fly ball." Sport is full of such lingo. All these home-made verbs take regular "-ed" endings in the past tense, even if a similar-sounding verb is irregular. Thus, the army didn't rang the town, it ringed it.

It is a distinction that people observe but don't think about.

When linguists ask, "Who threw the most underwear?" test subjects say, "She outflung him." When they ask who had the most affairs, the answer is, "She outflinged him." Children know the rule from the start, just as they know that irregular plurals can be used to make compound words but regular plurals (those formed with an "s") cannot. You can have teeth marks, but not claws marks.

Peter Gordon, a linguist at the University of Pittsburgh, tested three- and five-year-olds on this question. He said, "The monster that likes to eat mud is a . . . ," and the children said, "Mud-eater."

"The monster who likes to eat mice is a . . . "

"Mice-eater!"

"And the monster who likes to eat rats is a . . . "

"Rat-eater!"

The kids were just as reliable with nonsense nouns. But how could it be? Innatists decry the idea that the children could have learned it. After all, they point out, plural compounds appear only once in a million words. The rule is hard to explain, even to an adult. And no one is likely to correct a child who does make a mistake. The Chomskyites contend that the child must know all of these distinctions, and the thousands of others just as subtle, because the rules that govern them are innate.

"Now, I'm enough of a materialist to think that language is in the brain," Chomsky was explaining to the camera in room 250. "If you cut off someone's foot, he can still speak. In fact, it is useful to think of language as an organ of the mind. The brain is like every other system in the biological world: it has specialized structures with specialized functions, and language is one of these. But did we invent language because we were sentient? No more than we invented our circulatory system. What seems to be true about language is that its basic design is in the genes. The genes determine the structure and design of language. As far as we know, it is plausible to say that there is no variation in the computational system—in the principles that determine the organization of the series of noises that makes sense to us. All this happens in a very rigid manner, as rigid as the computation in your personal computer."

"No, no," Searchinger objected. "Would you start that again? It sounds too wordy."

Chomsky looked momentarily baffled.

"It's comparable to walking," Searchinger prompted him.

"Well, take, for example, the facility of walking," Chomsky went on. "If a child is raised by a bird, does he end up flying? No. Or if a dog is raised by a person, does it end up walking on its hind legs? No. That we are designed to walk is uncontroversial. That we are taught to walk is highly implausible."

Listening to the explanation unfold, I was reminded of why different disciplines have wished so fervently to keep hold of the language question: it is a hard one to divide up and share. Chomsky started out talking about language, and pretty soon he was talking about the nature of man. He had already gored a sacred precept—motherhood. According to Chomsky's innatists, children weren't learning language from their mothers, or from anyone else in their environment. They were bringing language with them.

The contention affronted common sense, and though it is now widely accepted, it still draws fire. "The innatists think that language is acquired very fast, very easily, and that it's very much the child's responsibility," Catherine Snow explained to me. "They also see language as one large problem. We on the other side think that learning language is a long slog, which requires from the child a lot of work. And the child is working as hard as he can, fifteen, sixteen hours a day. We think it requires a relationship with an adult, and a whole set of cognitive abilities. We also think that the child is refining one little bit of the language system at a time.

"People who are inclined to fall back on innatist explanations are falling back on a metaphor. It's an exciting metaphor. The image that transfixed them was that of the child as linguist: in his every utterance, he is the perfect speaker of an exotic, weird language. But even the most rabid innatist cannot point to a gene or a cell for language. And even the most rabid environmentalist must concede that language doesn't get learned by every species, and that if too much of the brain is missing you won't learn language. The solution lies somewhere in the middle. The problem is taking it out of the realm of mystery. The Princeton psycholinguist George Miller said, 'The trouble with language acquisition is that the nativists have proved that it's a mystery and the environmentalists have proved that it's impossible.'"

In the MIT lecture hall, Noam Chomsky and Gene Searchinger were finding it impossible to proceed with the filming: a scheduled class was arriving, and a professor had come in and nodded timidly in Chomsky's direction before turning and writing "Developing Amphibian Oöcytes" on the blackboard.

"Suppose that a child hears no language at all," Chomsky was saying. "There are two possibilities: he can have no language, or he can invent a new one. If you were to put prelinguistic children on an island, the chances are good that their language facility would soon produce a language. Maybe not in the first generation. And that when they did so, it would resemble the languages we know. You can't do the experiment, because you can't subject a child to that experience."

The lights flashed off, and the film crew began hurriedly packing up cables and microphones. "Of course," Chomsky commented to Searchinger as the two pushed against an incoming tide of undergraduates and headed for the MIT quad, "there are natural experiments."

8

The luck that befell Susan Curtiss when she was invited into the Genie case by Victoria Fromkin was greater than she at first knew, for the competition for access to Genie was fierce. Even by early May 1971, six months after the girl's discovery, there was no assurance that any linguists would be included among her scientific observers. And the scientists weren't the only ones trying to gain entry. "Immediately, there was such interest in Genie, such publicity," Howard Hansen, who was then the head of the psychiatry division of Childrens Hospital, told me. "We had calls from all over the world—press, doctors, do-gooders, kooks. We tried for anonymity. But we had to keep her in the hospital. She was a ward of the court at that point. If we had discharged her, she would have gone to Juvenile Hall, and that would not have been right. So David got active on a research design, and we put together a little money."

"David" was David Rigler, a professor of pediatrics and psychology at the University of Southern California and the chief psychologist in the hospital's psychiatry division. He had been with Childrens a year, having worked previously as an evaluator of grant applications for the National Institute of Mental Health, in Bethesda, Maryland. His experience proved useful in helping the hospital secure initial funding for research on Genie from two foundations and, in February 1971, a contract with the NIMH itself for $21,500. The NIMH contract would run until the following September, during which time a number of consultants were to be invited in for preliminary research and a conference was to be mounted to debate long-range plans.

Hansen and Rigler acted as gatekeepers for the process, with help from another hospital psychologist, James Kent. Kent's presence especially seemed to bode well for Genie. He was an authority on a phenomenon all too familiar now but not often acknowledged twenty years ago. Hansen elaborated: "I had been a pediatrician. Even back in the 1950s, I was used to seeing children's X rays of strange combinations of injuries, what we have come to call unassociated fractures: healing fractures of long bones with subdural fractures of the skull. We didn't know what we were seeing. We were just becoming aware of physical abuse of children, and not psychological, and certainly not sexual—that was still taboo."

James Kent knew what he was looking at. He was enough of an expert in child abuse that in 1972 he was appointed to a White House commission studying the syndrome and would become further convinced (though the Nixon White House, it turns out, eventually was not) that the nation was responding inadequately to a major problem. Kent was the doctor originally in charge of following Genie's case. "I was supposed to give Genie therapy," he recalls. "But mostly that entailed watching her improvement, documenting her progress. I became more her Boswell than her therapist."

The day after Genie's admission to Childrens Hospital, Kent paid her a visit. She had arrived in diapers and was having them changed when he walked in. After she had been successfully outfitted in a new set of pajamas, she got out of bed and shuffled toward him, apparently attracted by what he had brought with him: a magazine, drawing paper, crayons, and a Denver kit—a set of toys used to gauge the developmental level of young children. He was amazed at the skill with which she flipped through the magazine. It seemed that all her dexterity was in her fingertips, for tests were showing her to have, in general, the motor skills of a two-year-old. As Kent removed items from the Denver kit—a bell, a block, a small doll—she took each one and held it momentarily to her cheek but then laid it aside. She made good eye contact with him, seemed very curious about her environment, and was attentive to sounds, moving about the room to determine the source of each. This Kent found promising. But his overall assessment was bleak. "As far as I'm concerned, Genie was the most profoundly damaged child I've ever seen," he told me recently. "There has been nothing in other cases to approach it. It was orders of magnitude worse. Even the child I have cared for recently, a hearing child raised on sign language in a satanic cult surrounded with ritual murder and prostitution, all kinds of shit, is normal compared with Genie. Genie's life was a wasteland."

The question for Kent—and a question eventually relevant to Curtiss's work as well—involved what this damage meant for Genie's emotional and intellectual state. Because she couldn't talk, testing her intellect was almost impossible. But she was expressive of some emotion: Kent noticed her fear when he pulled a puppet from the Denver kit. Genie started, yanked the puppet from his hand, and threw it on the floor. Kent feigned a horrified concern and said, "We have to get him back." To his astonishment, the child repeated the word "back" and gave a shrill, nervous laugh. Encouraged, Kent began a slapstick pantomime, picking up the puppet and letting Genie throw it again, which she did with bursts of laughter. She was playing, and was quick to enjoy his reciprocating play.

She showed little beyond this, and Kent reported in a 1972 symposium paper that "apart from the peculiar laugh, frustration was the only other clear affective behavior we could discern." The frustration was just as peculiar. She would scowl, tear paper, or scratch objects with her fingernails. When she was very angry, she would scratch her face, blow her nose violently into her clothes, and urinate. But she would not make a sound, and she would not turn her anger outward, toward another person. It was not until later that her caretakers discovered how forcefully she had been coerced into suppressing all expression.

Kent continued to wonder at the child's hidden heritage. She had been "deprived and largely isolated," he noted in his symposium paper, "but she hadn't been asleep. She already had a store of experiences, memories, and expectations that must in some way color and shape her responses."

At first, though, she responded hardly at all. Her usual comportment, Kent noted, was a "somber detachment." If not deliberately engaged, she drifted around in her new physical world, walking with bent elbows in her strange "bunny walk," spitting into her clothing or into a curtain hem, far more aware of the room than of the people in it. In fact, she seemed hardly able to differentiate between various visitors. Some observers referred to her as "ghostlike."

9

Among the first of the outside consultants to visit Genie was Jay Shurley. "I invited Jay to come out," Hansen told me, "and I picked him up at the airport. I remember him getting off the plane, this big Texan. It was pretty funny. He had a cowboy hat and all."

"That first trip, I paid my own way," Shurley recalled recently. "I spent a week with Genie, examining her clinically. I determined for myself that she was the genuine article—that she had suffered the most extreme long-duration social isolation of any child that had been described in any literature I could find."

Shurley had sent the bulk of his luggage overland—six hundred pounds of state-of-the-art equipment for investigating brain activity. For three nights running, on each of his three early visits, he wired Genie to an array of meters, measuring her brain waves while she slept, looking for anomalies that would imply abnormal brain development. "Genie was about the richest source of information you can imagine," he said. "I responded to this, because I'm an investigator on a fundamental level. There were all kinds of questions that I felt she might shed light on. Naturalistic cases of intense isolation don't come along often—not with a period of isolation as extensive as that."

Shurley had a charter interest in the isolation question; he had grown up unusual, in a hardscrabble Texas farm family. "I was a black sheep," he told me. "My family are all ranchers. I'm the first one that wanted to go to college and become an academic." After graduating from the University of Texas Medical Branch, at Galveston, Shurley went to Pennsylvania Hospital, in Philadelphia, for his psychiatric training. After a brief stint in private practice in Austin (where he was, for a while, the only clinical psychiatrist in southern Texas), he was drafted into the army and stationed at Brooke Army Medical Center, in San Antonio, where he taught medical doctors who were accompanying the troops to Korea how to perform psychiatric services.

After this tour of duty, he moved to Bethesda to become chief of the Adult Psychiatric Branch of the NIMH. His official research concerned schizophrenics, and his fascination with their hallucinations led him to some controversial extracurricular experimentation. He spent his off-hours helping to develop the warm-water sensory-deprivation chambers that eventually made their way from science to parapsychology. Through the late 1950s and early 1960s, first at the NIMH and then at the Veterans Administration hospital in Oklahoma

City, Shurley used the tanks to experiment on himself, floating in their null environment until he experienced the vivid hallucinatory state of the disembodied mind. Some of these dream states reminded him of reports he had heard in the military—the accounts of test pilots who flew the new reconnaissance jets so high that they could see neither clouds nor horizon and so fast that they escaped the sound of their own engines. The air force denied that its pilots were hallucinating in flight, but the pilots themselves had a name for the point at which they seemed to depart from reality and enter the dream state—the breakoff. Similar dislocations were reported by soldiers stationed at lonely DEW-line outposts and by released American POWs returning from North Korea, where they had been kept in solitary confinement. Shurley realized that what he was experiencing in the tanks was really a combination of two phenomena, which he wished to tease apart. "You cannot achieve sensory isolation without social isolation," he explained. "For an intact, developed human being, the richest source of sensory contact is input from a fellow human being."

To study the effects of social isolation independent of the sensory, Shurley went to places where there were few human beings. He studied seamen on small ships, and in the sixties he spent three summers in Antarctica, recording the metabolism, sleep patterns, and psychosocial behavior of scientists and work crews sent there for thirteen-month stints by the National Science Foundation. He became such a fixture on that continent that the National Geodetic Survey named a mountain in the Pensacola Range Shurley Ridge. Students at the University of Oklahoma renamed his graduate course the Twenty-Foot Stare in the Ten-Foot Room. The equipment he hooked up to

Genie was stickered with bills of lading from the South Pole.

Of his first visit with the child, Shurley remembers that she treated everything, including people, as objects. "If you gave her a toy, she would reach out and touch it, hold it, caress it with her fingertips, as though she didn't trust her eyes," he told me. "She would rub it against her cheek to feel it. So when I met her and she began to notice me standing beside her bed, I held my hand out and she reached out and took my hand and carefully felt my thumb and fingers individually, and then put my hand against her cheek." His clinical experience provided a context for this odd behavior. "She was exactly like a blind child," he said. "She didn't integrate tactile and visual information. Even the bunny walk—hands in front. It's what we call a blindism. It's what people do when they do not entirely believe their eyes."

Shurley arrived on the scene in time to note some of Genie's initial progress. "When I saw her first, there was pendant flesh hanging around her buttocks where the hole of the potty seat had been. It was bruised black. There's no record of this except in my memory. Three weeks later, it had been reabsorbed, and the bruises had gone from blue to yellow." When he returned some two months later, he noted other, less encouraging transformations. "From being a totally neglected waif at the time I did my consultation, by the time I came back Genie had become a prize," he told me. "There was a contest about who was going to investigate her, and how—about where to go with the treatment and research. You can't go everywhere. There were several leads, and after my initial sleep study I was trying to figure out where *I* wanted to go. Language acquisition was part of what I was interested in, but not a predominant part. Victoria Fromkin had declared an interest in the cognitive area, but if Genie turned out to be a mentally retarded child—genetically or because of her diet—she wouldn't be a good case for study of cognitive development. The potential for cognitive development would not be there; there would not be a flowering. This girl had lived on gruel and on milk from nursing bottles.

"I thought it would be easy to investigate whether her brain had suffered deprivation nutritionally, informationally, socially. I wanted to know what the effect was on her growing brain and, secondly, on her growing personality. I was more interested in the socioemotional aspects than in the cognitive. An issue that I thought could really be explored was whether she could be reattached to a maternal figure. I thought it important to put her in contact with someone she could bond with. This case was something that was not duplicable. It was important that it be exploited fully and properly—and I don't mean exploitation in a pejorative sense."

Some of Shurley's ideas for handling the case seemed novel, if not extreme. In one meeting during one of his early visits, he suggested that Genie might benefit if she were placed back in a very restricted environment "highly similar to the one from which she emerged," except without the punishment and abuse. Then she could be brought out gradually, in a process more closely controlled and documented than her earlier helter-skelter emergence. This would correct any trauma from her sudden birth into the world, Shurley suggested. Of course, it would also provide a better baseline for the scientists' observations. The idea struck Rigler as an original one, "worthy of consideration," but he worried that Genie's decompression had already gone too far to humanely reverse. He turned the suggestion down.

Over the course of his several visits, Shurley's hopes for the research dimmed. "It was a politics-ridden situation, a matter of internecine warfare, almost from the word go," he said. "Childrens Hospital was an extraordinary location for pursuing a process that should be quiet and calm. Very soon, there was this breakdown—this conflict between doctor and hospital, between teacher, school, psychiatry, psychology. It became almost an armed camp, very quickly." Shurley attributes this breakdown to competition for credit in an institution afflicted with a "glitz factor" because of its proximity to the celebrity community. Those within the hospital, noting its gritty environs and unglamorous clientele, describe the internal politics as being more conventionally bureaucratic.

10

Genie, for one, seemed oblivious to the battles behind the scenes. For the first time in her life she was being treated relatively the same as other children and was, relatively, thriving. Her mental and physical development had begun almost immediately on her admission to the hospital. By her third day there, she was helping to dress herself and was voluntarily using the toilet, though her incontinence problems were to persist. After two weeks, she seemed ready for another expansion of her world and was released into the hospital's rehabilitation center, a single-story building with a yard and a play school, set apart from the hospital proper. There she was free to wander or watch, or to join in playing games and using arts and crafts materials alongside much younger patients. While the other children learned creative discipline, she learned freedom. She discovered that when she dropped things, even things that broke, she was not admonished and might, in fact, be encouraged to repeat the action. Her response to this license was what James Kent called "the most spontaneous and sustained" of her affective reactions.

"She entered quickly into a ritual play," he reported in his 1972 symposium paper, "during which she would eventually destroy the object. The nervous, tense laughter first associated with these episodes gradually changed to a relaxed and infectious laugh that would sometimes double her up and bring tears to her eyes. She would often accompany her own actions with cries of 'Stop it'—burst out laughing and repeat the action." Despite the disapproval of some on the staff who feared that Genie would go too far in this atmosphere of permissiveness (as she indeed seemed to do one day when she gleefully jumped all over her new eyeglasses and threw them onto the roof), Kent condoned her small orgies of destruction, seeing them as "attempts at active mastery of formerly traumatic situations."

Actions that would have earned a normal child a spanking seemed in Genie to be healthy signs of emergence. One day in early spring, she made hitting gestures at a new girl in the rehabilitation center, much to the surprise and pleasure of her observers. Previously, her rage had been directed inward. Susan Curtiss wrote in her dissertation: "Genie would erupt and have a raging tantrum, flailing about, scratching, spitting, blowing her nose, and frantically rubbing her face and hair with her own mucus, all the time trying to gouge or otherwise inflict pain on herself—all in silence. Unable to vocalize, Genie would use objects and parts of her body to make noise and help express her frenzy: a chair scratching against the floor, her fingers scratching against a balloon, furniture falling, objects thrown or slammed against other objects, her feet shuffling. These were Genie's noises during her sobless, silent tantrum. At long last, physically exhausted, her rage would subside, and Genie would silently return to her undemonstrative self."

Now, finally, Genie had turned some anger outward, aiming it at a source of frustration. She was upset with the new girl because she was wearing a dress from the hospital laundry that Genie had formerly worn; the episode was the first indication that Genie was developing a sense of self.

She already had a sense of possession; she hoarded found objects—books, paper cups, and anything made of plastic. Gradually, she showed signs of extending that possessiveness to people. From the start, her routine had included daily walks around the grounds with James Kent and, on most days, a drive with him to a local store or park. As was her habit, she seemed curious about him and glad to see him when he arrived but did not show in any way that she distinguished him from anyone else or mourned his absences. A month passed before a fleeting facial expression indicated that she registered his departures; finally, after another month, she reached over one day and took his hand to detain him. From then on, she would pull him back down to sit beside her when it was time for him to go. She cared not at all for other children; her attachments were to adults—especially to men who, like Shurley but unlike her father, wore beards.

She made friends with women as well—particularly with a woman named Jean Butler ("Miss Butler" to the children, a title Genie abbreviated to "Mibbi") who administered the special-education program at the rehabilitation center, under the aegis of the Los Angeles Public School District. "An extraordinarily dedicated teacher," Kent called Butler in one report, an estimation that Shurley expanded on. "Jean could identify with a child like *that!*" he told me, snapping his fingers. "She was a childless woman who loved children. She could empathize with them, even those who were the most withdrawn or handicapped—and there were some hard cases in the rehab center. Butler had been sick herself as a child, with vasculitis, a blood disease that had kept her out of school for whole semesters in her youth. That taught her compassion and patience, and she was capable of extreme patience with severely disturbed children."

Genie also befriended the center's handyman and a couple of the cooks, and it was to the latter that she turned early one morning when an earthquake hit Los Angeles. Running into the kitchen, she began verbalizing so profusely that one of the cooks commented later that if there had been one more tremor Genie would have achieved normal speech on the spot.

And she *was* achieving speech, if not quite on the spot. Her curiosity about her new surroundings sent her on a constant quest for the names of things. She would lead one or another of her caretakers around, using their fingers to touch or point to objects, while they said the corresponding words. "Hungry to learn the words for all the new items filling her senses,"

Susan Curtiss wrote, "she would at times point to the whole out-doors and become frustrated and angry when someone failed to immediately identify the particular object she was focused on."

Yet, although Genie's vocabulary increased, her speech stayed limited to a few short utterances; it soon became clear that she was understanding more than she could produce. During a class at the rehabilitation center one day in May, Jean Butler asked a boy who was holding a couple of balloons how many balloons he had. "Three," the child said, and Genie, looking startled, handed him the extra balloon he needed to make his answer correct. Intelligence tests were now being administered to her, and she was showing remarkable progress, gaining in some areas a year in development every few months. She showed what experts in child development refer to as scatter: on some skills—in the performance of such routine tasks as bathing herself, for instance—she scored the same as an average nine-year-old; on others, such as her almost complete inability to chew food, she scored as a toddler. Within the scatter, language remained near the bottom.

She was, at any rate, exceeding expectations, and in May her progress suddenly accelerated. Her vocabulary quest became more assertive; and her spontaneous (if largely incoherent) verbalizing, more frequent. She gained confidence in her movements, and began actively engaging in horseplay. She wanted to be carried piggyback, or to be swung around in the air like a whirligig. She was thrilled when someone holding her pretended to let her drop. "A great change from the child we saw at admission who shrank from most physical contact," Kent noted in his symposium paper.

11

May of 1971 was also decision time, when, under the terms of the NIMH contract, the consultants who had been observing Genie were scheduled to convene to consider her future. Several less formal meetings had been held, but this was the official one, on which the decisions about therapy and research and the application for a long-term grant would be based. David Rigler and Howard Hansen sent out the invitations; participants were booked into the Hollywood Plaza Hotel, on Vine Street. The first evening—Sunday, May 2—they were invited to Hansen's house "for drinks and chatter." The next morning, the chatter over, the discussion began in earnest, in the boardroom of Childrens Hospital.

The stakes were clearly high. From time to time, closet children (as imprisonment cases like Genie's have been called) and wild children (those abandoned as infants in the wilderness) have surfaced, and they have traditionally given rise to very visible science. Visible, difficult, and usually, in the long run, dubious.

The first feral child to come to the attention of what might be called modern science was Victor, the so-called Wild Boy of Aveyron, a pitiable creature discovered in January of 1800 lurking naked in front of a tanner's cottage in the Languedoc region of southern France. He was almost completely wild, having reached an age of approximately twelve in a state of independent savagery, living in the woods and eating acorns and pilfered potatoes. He had no language; his last human contact seemed to have been with whoever had cut his throat and left him to die when he was little more than a toddler. "Rescued," he was brought to Paris, to the Institut National des Sourds-Muets, there to be observed, taught, tormented, and loved by a young physician named Jean-Marc-Gaspard Itard. So varied and fruitful was Itard's career that it gives an impression of professional profligacy; he has been called, among other things, the father of child psychology and the father of the study of ear, nose, and throat disorders. Victor was his most celebrated and most frustrating subject.

The emotional connection between the ambitious teacher and his strange student is apparent from Itard's notes. He tells of the remorse he felt when his pressuring induced quiet tears or sobbing tantrums, of how he would sit immobile for minutes while Victor sat before him fondly caressing and kissing the teacher's knees. Even so, Itard could not refrain from using the boy's affection as a tool—challenging his trust by terrorizing him with a Leyden jar (a sort of battery that can deliver a shock) and unfairly punishing him over his lessons to test his sense of justice. Victor knew enough about justice to be outraged, and Itard found the outrage edifying. Under Itard's aggressive instruction (he once dangled the boy from a fifth-story window to frighten him out of his recalcitrance), Victor made some hard-won headway. He learned to spell the French word for milk, and on visits to a neighbor's home would take along the appropriate letters from the institute's metal teaching alphabet so that he could spell out "L-A-I-T" while downing a glass of it. But he never learned to talk.

Victor was nonetheless influential. In 1912, the Italian educator Maria Montessori called Itard's work "practically the first attempts at experimental psychology," and she based some of her innovations on his experience with Victor. The metal cutouts of letters and shapes still common in Montessori classrooms are descendants of the ones that Victor used. In other ways, too, the world is different for Victor's having come under scientific scrutiny by people who understood methodology and the merits of objective observation. Even so—as Thierry Gineste, the reigning expert on the Wild Boy, contends in his book *Victor de l'Aveyron: Dernier enfant sauvage, premier enfant fou*—the useful knowledge arising from the case was limited by how little was learned about the boy's past and about his potential. He remained, in the end, an enigma.

Among the cases of wild children discovered over the last seven centuries, more than fifty have been documented. The list includes the Hesse wolf-child; the Irish sheep-child; Kaspar Hauser; the first Lithuanian bear-child; Peter of Hanover; the second Lithuanian bear-child; the third; the Karpfen bear-girl; Tomko of Zips; the Salzburg sow-girl; Clemens, the Overdyke pig-child; Dina Sanichar of Sekandra; the Indian panther-child; the Justedal snow-hen; the Mauretanian gazelle-child; the Teheran ape-child; Lucas, the South African baboon-child; and Edith of Ohio. Investigations of these cases were generally marred by an excess of enthusiasm and a lack of methodology on the part of those who could have turned the children's misfortunes into revelation; by Genie's advent, a sorry pattern of missed opportunities had been established. "When an experiment like this comes along, there is intense excitement, and intense pressure," Jay Shurley remarked to me. "People tend to operate in these situations much more with their thalamus than with their cortex."

On the first day of the conference in Childrens Hospital, Shurley gave the results of his sleep studies. Genie's brain waves, he said, had shown a large number of what are called sleep spindles—artifacts that may indicate retardation. Others' observations were more subjective, less technical. Jean Butler reported that Genie was euphoric on holidays and weekends, when she got to leave the rehabilitation center on chaperoned trips; that she often said "No" but didn't mean it; that she called people "peepa"; that "dert" meant "doctor." She had had no problem with urine soiling since Christmas. She had been afraid of some boys who one day came past the classroom windows carrying rifles. She was terrified of big dogs, and of all men wearing khakis. She thought that singing was exclusively for her benefit. Videotapes were shown of Genie in the rehabilitation center, and Rigler described a party that had been held there to celebrate her fourteenth birthday. It had overwhelmed her, he said, and her anxiety had mounted with each present opened, until at last she had to leave the room and sit in a corner holding Rigler's hand while she calmed down.

The second day was reserved for "deliberations of the consultant panel," meaning that it did not include those people seen only as caretakers, like Butler and the rehabilitation center cooks, who had been invited to participate on Monday. ("So Genie responds well to your intrasupportive initiatives?" a scientist had asked one of the cooks. "I just gives her love," the cook had replied.) Tuesday was for scientists only; besides

Shurley, Rigler, Hansen, Kent, and Fromkin, there were some fifteen psychologists and neurologists from all over the country. When they convened, their discussion was shaped as much by an event of the previous evening as by the first day's testimony.

12

It is one of the resonant curiosities of Genie's story that her discovery coincided with the Los Angeles premiere of François Truffaut's *The Wild Child,* a movie that tells the story of Itard and Victor, "*l'enfant sauvage de l'Aveyron.*" Between the newspaper accounts of Genie's rescue on page 1 and the cinema ads in the entertainment section, art and life seemed to be doing a do-si-do.

"Have you seen the movie the WILD CHILD??" Jean Butler wrote to Jay Shurley in January. "I have not as yet. . . . So many friends who have seen it say that there are so many striking similarities between Victor and Genie. . . . Many felt it was a discouraging movie but all agreed that it followed Itard's case study to the letter. . . . In fact Truffaut reads directly from Itard's notes. . . . The film played at a theatre nearest the hospital for twelve weeks. . . . The theatre is owned by a friend of mine . . . he said it would be a privilege to show it privately during the day and to any of my friends who have not seen it or would like to see it again. . . . He needs only two or three days notice to show it privately and would do it at the Los Feliz Theatre which is just a few blocks from the hospital. . . . I'll tell Rigler and Hansen and let them arrange something."

The unguarded enthusiasm of Butler's letter (written on a typewriter given her by her friend the theater owner) is remarkable, coming from a woman whose later correspondence on the subject of Genie is notable for its aggrieved sobriety. "You undoubtedly heard about our earth quaking experience," she continued, and followed the sentence with six exclamation points. "I'll tell you about Genie's reactions when you are here. . . . She was terrified. . . .

"David Rigler is handling the aforementioned (in last letter) problems with the School district. . . . He is doing it with such diplomacy, I'm sure that someday we are going to lose this man to the State Department. . . ." Butler closed, "P.S. For the sake of science, keep your beard," and added for good measure four more exclamation points.

So at four-thirty on Monday afternoon, May 3, after the day's testimony on Genie was finished, the symposium members adjourned to the Los Feliz theater for a private screening of *The Wild Child*.

"The impact on the whole group was stunning," Shurley recalled. "At first, there was silence. It was very moving—no one could say anything. Once people overcame the shock, the questions began to flow." The questions flowed through dinner and into the next morning's session, but anyone who may have hoped that the film would promote accord among the attendees was quickly disabused. "There were so many things commented on," Shurley said. "All of us saw in the movie what we were prepared to see to confirm our own biases."

The biases concerned two areas: what Genie could best reveal to science and what, in the course of that revealing, science could ethically ask of Genie. Shurley's handwritten notes of the Tuesday meeting include the sentence, "Rigler talked on second day on constraints on research, legal and moral." After the movie, even more than before, moral concerns seemed to be on everyone's mind.

"My pitch was—and some others agreed—that the interests of the girl, in terms of therapy, would have to be uppermost, and that anything we might learn from her should be a secondary consideration, and should be done within the context of her therapy," Shurley told me. "Others said that this was too great a scientific opportunity—that research had to be primary."

Three months after the conference, Rigler elegantly expressed the interdependence of the two themes in a letter to Jean Butler. "Justification for these [NIMH] funds was the scientific importance associated with the study of this child, study that was based essentially upon successful rehabilitation," he wrote. "Theories of child development hold that there are essential experiences for achievement of normal psychological and physical growth. If this child can be assisted to develop in cognitive, linguistic and social, and other areas, this provides useful information regarding the critical role of early experience which is of potential benefit to other deprived children. The research interest inherently rests upon successful achievement of rehabilitative efforts. The research goals thus coincide with [Genie's] own welfare and happiness. Conversely, if our research methods were to interfere with [her] development, they would defeat the very purpose of the research."

In Shurley's recollection of the conference, science was already interfering. "Dr. Rigler and others argued for the primacy of research—couched, of course, in ethically sensitive terms," he told me.

The meeting ended in what one conferee called "some considerable confusion." Rigler was left with the chore of digesting all the debate and deciding the nature of the final NIMH grant proposal—what kind of work the grant should fund and who should do it. The advice he had received was, perhaps, more than he had bargained for. "He looked like a man who's thirsty for a sip of water and is handed a fire hose," Shurley recalled. In a postconference letter, Rigler and Hansen thanked the conferees for an "enriching exchange" and solicited their reactions to the proceedings.

Those reactions soon rolled in, and some had a warning tone. David Elkind, a professor of psychology at the University of Rochester, wrote, "Although language is not my area, I would like to reinforce the words of caution I expressed at the meeting. Too much emphasis on language could be detrimental if the child came to feel that love, attention, and acceptance were primarily dependent upon her speech."

David A. Freedman, a professor of psychiatry at Baylor College of Medicine, in Houston, argued that the acquisition of speech might be dependent on what he, like Elkind and the cook, called love. He rejoiced in the evidence of Genie's progress presented in the videotapes, noting the "very dramatic... change in her appearance from apathy, to a wan and pitiable appearance, to an at times animated and involved little girl, which seemed to correlate with the passage of time." But his clinical experience with other unfortunate children had taught him to be cautious of the varnish that videotape and optimism can apply to such cases. He was unconvinced by surfaces. He was looking for a thaw at the center, and a visit he had had with Genie had disquieted him:

When I arrived she was having her breakfast. Although she sat at the table with two other children who were engaged in fairly typical childish conversation and play, she had nothing to do with them. It is difficult to put into words the feeling I had about what she did. I don't think it would be accurate to say she actively ignored or rejected them. Rather it seemed to me that it was as though for her they were no different from the walls and furniture in the room. . . . The question becomes how to go about inducing in this child the ability to be aware of both herself and others and feel an interest in and need for others. My prejudices say that if this goal can be achieved she stands a chance of leading a relatively normal life; if [it] can't, she will remain an automaton. My prejudices also say that to achieve this goal it will be necessary for Genie to establish a particularly close relation with some one person whose care for her will include the provision of a good deal of body pleasure. I'm referring to something analogous to what any good mother automatically and unconsciously provides her infant as she bathes, feeds, and diapers it. Obviously this won't be easy to do for a fourteen-year-old. Yet, I believe a necessary precursor to any effective educative process would be her development of an intense, dependent attachment to some one person whom she would be interested both in identifying herself with and pleasing. . . .

Without the creation of such an attachment, and all it implies with regard to Genie's need to attempt to maintain it, I doubt whether she will have the equipment to integrate whatever skills she develops. I believe something along this line was implicit in the sense of the group when we were all in accord that it would not be indicated to attempt to train Genie in talking. . . . She should be, in my view, bathed, clothed, toileted, massaged, kissed, cuddled, and fondled all by one person. Other people should be available but in a distinctly secondary role. Out of such an intense relation should grow both an awareness of herself and of whoever it is who is caring for her. Such an awareness, to reiterate, seems to me to be the necessary first step in her education.

Later that summer, Rigler made his grant decision, and its primary focus was on language acquisition—not teaching Genie language so much as watching how she learned it. The main beneficiary was a scientist whom Shurley had taken little notice of during the May session. "It was a surprise when I learned that Victoria Fromkin initiated a major study," Shurley told me. "But Rigler thought a language study was a good idea."

Shurley fully understood why deciding the direction of the case might be a perplexing chore. "At first, confronted with this child, we didn't know what questions to ask her," he told me. "Genie was an absolutely beautiful example of a process: when confronted with nature—human nature—in the raw, you stumble around and come up with one or two questions to ask. The questions come out of your culture. The Wild Boy of Aveyron—Victor—came along when all the questions of the Enlightenment were being asked, and they were asked of him. But he didn't answer them. If you choose the right questions—which is to say, if they are the relevant questions—then you get around to the content, and you begin to read what was written there all along."

II

PREMONITIONS

13

Struck though Shurley and the other conferees were by Truffaut's movie, they could not have imagined, as they sat in the otherwise empty Los Feliz theater, how deep would run the parallels between the two so distant cases—the boy abandoned to the forests of revolutionary France and the girl trapped in a twentieth-century American suburban bedroom—or how insistently the similarities would surface. Indeed, simply by viewing the movie the committee was aligning the case at hand with the one on the screen: in 1800, the scientists deciding the fate of the Wild Boy had also sought counsel from popular entertainment. They had attended a play, then the rage in Paris, about a fictitious *enfant sauvage*. The melodrama was called *The Forest's Child,* and Victor was named after its protagonist.

Like Genie, Victor seemed on discovery to be impervious to heat and cold: he pulled potatoes out of the fire with his bare hands, and he cavorted naked in the snow. (Later, even after Itard had admired Victor's first sneeze as an indication the boy was being acculturated to a certain sensitivity, Victor would persist in rushing barely clothed into the gardens of the Institut National des Sourds-Muets to luxuriate in an unexpected snowfall.) Like Genie, he seemed not to make distinctions between what could best be perceived by feel and what by sight, suffering from what one attending scientist termed "a dissonance of vision and touch." Like Genie, he was also substantially oblivious to the existence of anyone but himself. ("I am dismayed to see the natural man so egotistical," reported J.-J. Virey, one of Victor's first observers.) As would be the case more than a century and a half later, the egotism seemed, at least on the surface, gradually to melt. Like Genie in the rehabilitation center, Victor adopted as a favorite activity the setting of the table. One day, he set a place for the just deceased husband of his loving caretaker, Mme Guérin, and her tears astonished him; it was his first encounter with human grief. He put the place setting in the cupboard and never brought it out again.

As with Genie, Victor's discovery occasioned a sideshow, though on something of a grander scale. His arrival in Paris from the departmental capital of Rodez—the trip, by coach, had taken a week, during which the boy was kept on a leash—created a public furor. Rumors flowed through the crowd surrounding the institute grounds that he was perhaps the long-lost Louis XVII, who, like some premonitory Anastasia, had survived the execution of his royal parents and was said to have fled into the forest; however, the foundling's age seemed wrong. A new play supplanted *The Forest's Child* in the Paris theater vogue, a vaudeville number written specifically about the Wild Boy and called *The Savage of Aveyron* or *Don't Swear by Anything*. On the ancient rue Saint-Jacques and its crooked tributary, rue de l'Abbé de l'Epée, oddsmakers set up shop, taking bets on whether the boy would ever talk, ever be civilized. Newspapers carried the betting charts. Itard sequestered Victor from the more indiscriminate attentions; later, however, he acted as Victor's chaperon among the perils of Parisian high society. When the two accepted a summons to dine with Mme Récamier, the ravishing young socialite whose attentions conferred social beatification in the capital, Victor left the table and ran into the yard, tore off his clothes, and climbed a tree; he was not invited back. On another occasion, he met the Marquis de Sade—an encounter that the official history of the institute describes as "*vraiment un rendez-vous manqué.*"

The public's interest in Victor was not just morbid. Modern children who are abused or neglected draw our attention because we see them, usually, as disturbing exceptions, albeit symptomatic ones, to society's prevailing order. In France in 1800, order was not presumed; the Committee of Public Safety and the Reign of Terror had taken care of that. Even in the prevailing order of earlier, calmer eras, children did not enjoy their current cosseted status. The Enlightenment's emphasis on the worth of the individual had been extended to individual children, but in a grudging sort of way, and the expedience of leaving them—at least, the unwanted ones—to die in the woods was not unheard of and not altogether shocking. The boy found naked in the tanner's doorway was interesting to his country's citizens not because his brutal history astonished their finer sentiments but because the Enlightenment and the Terror had honed an appreciation of certain questions that the boy might be able to address—questions about the nature of man. Strange as it seems in an age in which philosophy is to pop culture a thing apart, the betting sheets in the *journaux* of Paris were a street referendum on the ideas of

Montaigne, Rousseau, Descartes, Condillac, and Locke.

Whatever its more general effects, the Revolution seemed to have worked to Victor's advantage. Foremost among its courtesies was its timely end, which permitted a renewal of interest in things scientific. During the preceding decade, Paris had not been a happy place for scientists, among others. Intellectual independence had been considered almost as subversive as priestly piety. The Society of Observers of Man, the anthropological organization that initiated the research on Victor, was only a month old when he was discovered. Ten years earlier, the revolutionary government had sanctioned the institute where he was to live, adding "National" to its name and supporting it from state coffers. Before the school's successful efforts to teach sign language to the deaf, they had been considered subhuman and had been locked away in the purgatory of the Bicêtre asylum, with criminals, epileptics, and the insane. For the new government, the ability of deaf people to communicate was a symbolic resurrection, a metaphorical promise to the voiceless of all kinds.

Instead of delegating the usual policeman to run the Bicêtre, the government had appointed a doctor, Philippe Pinel, who would become known as the father of psychiatry. Like Abbé Sicard, the director of the Institut National, Pinel played a role in Victor's education; the two proclaimed him unsalvageable, a "*prétendu sauvage*" but a true and irremediable idiot. After that harsh dismissal, the boy languished for months in a limbo of neglect, until Itard, disagreeing with his mentor Sicard, took on the task of proving Victor's potential.

Itard had found contradictory revelations in the revolutionary spectacle. While the mad in the Bicêtre were ascending to humanity, humanity at large was descending willfully into madness. To Itard the convergence suggested a principle that has since been one of the bedrock assumptions of his profession—that the psychologies of monstrosity and of normality are closely allied, thinly separated. The idea pervades our culture. When we seek our warped reflection in the acts of Joel Steinbergs and Hedda Nussbaums, we rehearse that insight. In terms of intellectual provenance, the procession of gawkers parading past Genie's childhood home in November 1970, had departed from the vicinity of the Bicêtre and the walled gardens of the Institut National des Sourds-Muets, in the era of Revolutionary France. In other ways, too, Genie was awaited by the age.

Like bronze, French science is a useful amalgam of two slightly softer elements. Descartes set out the basic scientific method, rooting it in a rigorous adherence to what can logically be inferred; he trusted the corporeal senses the way a bedouin trusts the shimmer of silver in distant sands. A century later, the philosopher Étienne Bonnot de Condillac adopted a more generous opinion of outward experience. Taking his cue from the empiricism of John Locke, Condillac contended that our minds are blank slates at birth and are tutored entirely by our surroundings. The world lived in Descartes; Condillac lived in the world.

Their contest survives in the enmities of modern child development. Would Genie's success at language depend on her native resources, or on the ability of the world to reform her with its teachings? If she failed, would it be blamed on her biological clock, or on the ravages of her early experience? Descartes and Condillac continue to pick their fights.

Happily Condillac's empiricism was ascendant in France in the late 1700s—happily, that is, for an age in love with experimentation and exploration. Much scientific endeavor of the eighteenth century was aimed at determining the physical distinctions between man and beast. It had long been held by some that the physiological feature most innately human was the fanny, or perhaps the calf—or, at least, the upright posture that had created them. But then the voyages of exploration reached Borneo, where Europeans encountered upright and eminently fannied orangutans, and the distinction collapsed. (For some reason, it provided no comfort that humans were the only ones to make ships and go voyaging. If the orangutans had disembarked, upright, onto the shores of France, theirs would seemingly have been a kinship more worth worrying about.) Articulation of vocal sounds was another promising criterion, except that magpies could also do it pretty well, and New World parrots marvelously. And the ability to express emotions was the property of any pet.

So hotly contested was the border between men and animals that some unlikely incursions were made from either side. The incursions are instructive. In the sixteenth century, when menkind were still menkind, animals merely animate, and God still knew the difference, the question was whether the Indians discovered in the West by Columbus should be admitted into the family. It was only with the Papal Bull of 1537 that they were conclusively decreed as human and thus deemed worthy of conversion to Christianity. Two hundred years later, feral children were still similarly disfranchised—Linnaeus, in his *Systema Naturae,* named them as a separate species, *Homo ferus,* which he listed among the quadrupeds. Meanwhile, the orang-utan's possible humanity was being so seriously contemplated that some eighteenth-century scientists suggested mating one with a prostitute, to see what progeny would ensue.

Clearly, some defining event was needed. The scientists of Condillac's time, like physical anthropologists of a later day, sought a missing link—in this case, a living one, someone or something perched squarely on the species' frontier. By their orthodoxy, that would have to be either a talking ape or a human being who had been reared without human contact, like an animal in the wild.

14

Victor, even before he reached Paris, was debunking some of the cherished theories. To the dismay of the upright-stance advocates, he was seen, during one of his several escape attempts, to cross a field on all fours, running close to the ground, like an animal. J.-J. Virey found no sign of another "innate" human trait: "Is our young Aveyronnais capable of pity?" he asked. "Personally, I venture to believe that if this young man could bring some interest to bear on the things around him, then he would be inclined to commiserate as much as children ordinarily are." Like Genie, Victor hoarded what he cherished, and refused to share. Like Genie, he warmed only slowly to adults and not at all to other children. Having given the lie to physical rectitude and empathic feeling as defining characteristics, the boy, like Genie, was called to preside over a grander mystery—the mystery near the center of the web.

Montaigne said, in an essay of 1580, "I believe that a child brought up in complete solitude, far from all intercourse (which would be a difficult experiment to carry out), would have some kind of speech to express his ideas," and he implied that the inherent enigma was still that of Psamtik: Which language would the child speak? The Enlightenment tortured new subtleties out of that question. Was our native language that of the soul, or of society, or of the intellect? Did thought lead to language, and language to society? Some inverted the progression: society was our most innate characteristic, they said; it enabled language, and language enabled thought. Did the child in the woods not think, then? Was it possible to think with something other than language? Was it impossible to think alone? Or was thinking alone the necessary precursor to all else? The questions outlived the age. By the end of the nineteenth century, the German philologist Heymann Steinthal had concluded that language was not meant solely for communication. "Language is self-awareness," he said. "That is, understanding oneself as one is understood by another. One understands oneself: that is the beginning of language."

For Victor, all this distilled into a make-or-break equation: no matter whether he crawled or crept, if he could talk he would be judged, by humans, human. The equation was different, but hardly less compelling, for Jean-Marc-Gaspard Itard. If he could resurrect the boy from savagery, he would provide what he termed "concrete proof" of Condillac's theories. He would demonstrate that man brings nothing with him, that education is all.

However, for the young teacher and his young charge the beginnings of language were difficult to locate. In the drafty apartments of the Institut National, the two suffered together through one or another draconian teaching scheme for two years before Itard finally developed a system that showed some promise. He trained the boy to recognize certain written words and to connect those words with individual objects—the word *chaussure,* for instance, with a particular shoe. This accomplishment led to a game—a combination of flash cards and hide-and-seek—in which Itard wrote a word and Victor ran around their chambers seeking its correlate. Then Itard took the game a step further, depriving Victor of the specific shoe and making him seek others, thus forcing him to form a generalized notion of the word's meaning. For a while, the boy was off on a rocket ride of comprehension. He learned not only to find an object if he was presented with its written name but also to write the name when he was shown the object. And not just objects: he learned adjectives and verbs as well, with which he could both comprehend and concoct written sentences. Even this little bit of language seemed to open up new ways of thinking for him. The boy who had been completely adrift could concentrate. Chores he had performed mechanically were suddenly imbued with spontaneity and imagination. He even seemed better able to imagine the needs of others.

"All this went like a house afire," Roger Shattuck, a French scholar and translator, reports in *The Forbidden Experiment,* perhaps the most lucid of the English-language books on the Wild Boy of Aveyron. "Language now vibrated in all his actions."

In the end, however, it did not vibrate quite enough. The triumphant note at the end of Truffaut's movie marks the point of Itard's first report, presented in 1801, when Victor had made a certain amount of frail early progress and seemed on the verge of much more. Five years later, Itard offered the Society of Observers of Man his Second Report, and it is markedly different from the first. There had been progress, true, but Itard had come to appreciate the limits, rather than the potential, of his young student's mind. The boy was clearly capable of hearing and producing the necessary range of sounds, but he had shown that he would never learn to speak. His writing skills could proceed only so far. And his progress had been obstructed by the debilitating "crisis" of puberty, which drove the boy into torments and distractions that he was even less able to control or understand than other boys his age. Itard bled him to relieve his hormonal storms and recommended stopping the experiment.

The cessation presented a personal and professional postscript to the parable of teacher and student, and it was not a reassuring one. How much different would Genie's story have been if the movie that, in Jean Butler's words, "followed Itard's case study to the letter," had followed it to the end? How differently would the "profoundly affected" scientists have felt about their viewing of Truffaut's film if he had pursued his subject through its early miraculous optimism and into more cautionary territory? When Shattuck criticized the "amputated" nature of the film (a film he lauded as "an intelligent tribute" and "a nearly definitive account in the public mind"), Truffaut wrote him a response. "I don't regret having ended the film on a hopeful note, since Victor had actually learned a number of things and words," Truffaut said. "But I willingly admit that after the last scene, exactly as I did in *The Story of Adèle H.*, I should have added an illustrated narrative of what happened to the characters after we leave them on the screen."

After the characters were left on the screen, Itard became only more successful and famous; Victor, more forgotten. In 1814, Itard received the Medal of the Legion of Honor, and in 1821 he was elected to the Academy of Medicine. He spent his idler moments in a mansion in the Parc de Beauséjour, in Passy, the neighborhood of Rossini and Beaumarchais. After 1811, Victor lived with Mme Guérin, supported with a small state pension, in a small house near the institute, four doors down the Impasse de Feuillantines from his only neighbor who would ever encounter fame, the adolescent and yet impoverished Victor Hugo.

On Paris's rue Saint-Jacques, a good ways south of the Sorbonne and just before the street veers eastward toward the chapel yard of the Val-de-Grâce, there is a high masonry wall sheer to the sidewalk, with a tall, heavy door in its center. On a cold December morning I stepped through that door and into the cobblestone court of what is now named the Institut National des Jeunes Sourds. I had an appointment with a Docteur Karakostas, a psychiatrist in Paris who devotes Wednesdays to his role as the institute's archivist. The building's facade is little changed from the engravings of it made during Victor's time—five unadorned stories, shuttered casement windows.

Inside and out, the building presents a peculiarly European (and decidedly un-American) model of an eminent institution maintained in dignified, unglamorous austerity. The institute's floors are swept but not varnished. Its library contains a wealth of important documents, including Itard's hand-penned manuscripts pertaining to the birth of psychiatry and education of the deaf, but there is no staff librarian to oversee it.

Karakostas met me in the court and led me in through some glass-paned doors and up a wide curving stairwell to the top floor. His office is a table set between two banks of lockers behind a room where women in white smocks tend to the mending of the institute's laundry.

"The institute is in the center of the history of deaf education," he instructed me when we had found enough chairs that we could both sit down. Then, to the gentle percussive whirring of the sewing machines, he led into the small part of the institute's history associated with its most famous inmate.

"You can't separate the enthusiasm for the Wild Boy of Aveyron in 1800 from the social changes of the French Revolution, the tremendous interest in anthropology, the whole . . ." —and as he searched for his word, he held his hands before him as though over the ears of an imaginary recalcitrant head and shook it vigorously—". . . *brûlantment,*" he blurted finally, "of the epoque. This was at the beginning of psychology. The beginning of medical specialties. Itard founded psychiatry for children. He created a domain of study for otorhinolaryngology. He was the first to make the distinction between deafness and autism.

"It is fantastic to see how intense the relationship was between Itard and Victor," Karakostas said. "There was an intense will of knowing. It was a love story. Not a sexual love story, but pedophilic, definitely. The Wild Boy was to be observed, to reveal the moral ideas of the time, the big questions of humanity." Karakostas held up a fist and began itemizing, finger by finger, beginning, in the manner of the French, with his thumb. "Questions of language, psychology, education, the limits between animality and humanity, the relationship between deafness and cognition.

"But," he said, "he was a contrary example." And Karakostas leaned intently forward, his sentences suddenly clipped. "The Wild Boy never spoke. The Wild Boy was abandoned. The Wild Boy was put in a small house near here and eventually forgotten. In 1828, when he was about forty years old, he died. When he was put in that house, eight years or nine years—anyway, fewer than ten years—after he was found, no one cared. Itard went on working here. He did not care about Victor, and never saw him.

"The lack of interest," he said, and shrugged himself back into his chair. "It was not because Victor was unsuccessful, it was because the questions of the time had changed. The questions of the *Lumière* had gone underground. When a new wild child was found in the provinces, some years later, the government in Paris was notified," Karakostas said. "And the people in Paris said, 'You keep him.'"

Outside the institute, the second snow of the season was starting to fall, and I could see it through Docteur Karakostas's small, low square of window, gathering across the narrow street in the carved stone gothicry of the Church of Saint Jacques de Haut Pas. When the archivist was called away by students to open up the library, I let myself down the stairs and out the back of the building and took a walk through the walled gardens. Except for the modern annex under construction to the garden's rear, it was an acre seemingly undisturbed from its eighteenth-century inception, from the years when just such a snow as this would have pulled Victor bursting through the same glass doors. The deaf children of the institute tend this garden, now as then, but in this cold weather they were nowhere to be seen. The grapevines of a vest-pocket vineyard were trained up their trellises into gnarled candelabras. The central formal fountain, so prominent in old engravings, lay half full in its winter disuse, its surface crusted with old ice, powdered with fresh snow.

After Victor was discovered and before he ever saw Paris or this place, his provincial overseer, Bonnaterre, had felt moved to mourn the boy's inevitable future. Bonnaterre must have known that in some ways salvation from so brutal an abandonment would in itself be wrenching. "How utterly will you lose your independence, bound with our political shackles, caught in our civil institutions; you should truly weep!" Bonnaterre wrote. "The path of your education will be sprinkled with your tears."

The snow whirled nimbly over the ice, fell thick through the brittle air, and settled—startling white amidst the darkest red I may ever have seen—between the petals of the students' roses, blooming unmindful of season around the fountain's edge.

15

If the questions of the Enlightenment went underground, they didn't go far and have been utterly unruly about staying there. Their incidence is malarial, herpetic. Just when you think you've moved on to more modern perils in the Age of Deconstruction, here they come, recurring. When Noam Chomsky professes the innate nature of language, citing the inadequacy of the input the child receives from its encompassing world, and when Catherine Snow responds that she is sure the child must glean most of its language from its surroundings, they are donning Cartesian and Lockean robes. Genie intruded into that argument and fell into a wonderland of ancient rivalries. Her Hansens and Kents were children of Pinel, her Jean Butlers descendants of Itard. Condillac attended, his ghost guiding those who hoped that education would determine the remainder of Genie's life. Condillac is the patron, and Descartes the hobgoblin, of social workers everywhere.

Unlike most of the known wild children, both Victor of Aveyron and Genie of Temple City arrived to expectant audiences. Victor's debut was timed roughly to the questions of Condillac and precisely to the creation of the Society of Observers of Man. In 1971, Genie had the services of a different advance team. As David Elkind, one of her early observers, puts it, "Chomsky was new then, and linguistics was hot—there was a new theory coming out every day." Her arrival was even more precisely timed to the advent of one of those theories.

The study of language acquisition in children turns on a single simple idea—one that I heard most succinctly expressed in the keynote speech at the 1989 Stanford Child Language Research Forum. For those outside the profession, the Stanford conference and others like it can be strange affairs. Linguistics, like all ghettos, has its own internal language. Gene Searchinger's secretary once gave up in exasperation her attempts to transcribe a taped interview with a linguist, complaining to the filmmaker, "This person's using a foreign language to speak English!"

Not surprisingly, Linguese is at its most virulent where hundreds of linguists have gathered together; their professional conclaves can resemble a late night at a camp meeting where the faithful have lapsed into tongues. The Stanford Child Language Research Forum was the type of place where people argue over cocktails about whether children do or don't start with a trochaic bias in the word stream, where a hiss might be called a fricative, a kiss an ingressive plosive. It's the type of place where one woman could look over at her friend's two-year-old daughter cavorting in her new red dress near the registration desk and say to the friend in that hushed, motherly mixture of admiration and concern, "I see little Jessica still hasn't gotten her labials."

Around the glossary of hard-edged technical terms lies a soft halo of argot. "Story" in Linguese means a proposition, hypothesis, conceptual advance, scenario, explanation, or remarkable new fact. A good research paper is called a "very nice story." "Star" means "wrong." A grammatically poor sentence whose transgressions are revelatory is referred to as a "fact."

Over the breakfast table in her L.A. home one morning, Victoria Fromkin conspired with Penn's Lila Gleitman to give me an example of contentious Linguese:

GLEITMAN: "Linguists argue like this—'I believe in "He being a fool." Does that go through for you?'"
FROMKIN: "Star, star, star."

For a linguist, Gleitman has an odd predilection for English, and, in her Stanford keynote speech, she was characteristically plainspoken. In her late fifties, with close-cropped dark-gray hair, wearing sneakers, slacks, and an orange-patterned blouse, she managed to give the impression, as she leaned on the lectern, of a truant leaning against a gymnasium wall smoking a cigarette instead of going to class, and being too cool to care. "Can you hear me?" she barked into the microphone, and then snorted to herself, "Huh! Only too well." The snort, it turned out, was a trademark—the nasal harrumph of a prize-fighter, equal parts cynicism and deviant relish. On the movie screen behind her appeared a slide of the front page of a supermarket tabloid, with a headline reading "MOM GIVES BIRTH TO 2-YEAR-OLD BABY," beneath which was the subhead "CHILD WALKS, TALKS IN 3 DAYS."

The audience laughed. The speaker finished arranging her papers and looked up. "As by now you probably know, I'm Lila Gleitman," she said. "And basically what I want to talk about is this." She walked over and hit the screen a sharp one with a pointer. "What took three days?"

"What Took Three Days?" has been Gleitman's obsession for the last several decades, during which she has adopted, rather despite herself, an ardent Chomskian viewpoint. "People say, 'That Lila, she's just this crazy rationalist,'" Gleitman told me over lunch the day after her speech. "'She thinks everything's innate.' But I started out as a hard-core empiricist, honest! I designed my studies to prove the empiricist position, and I couldn't ignore it when they showed me to be wrong."

One of the experiments she designed was directly inspired by empiricism's patron saint. "Locke said, 'Look at blind people—there should be some things they can't learn,'" she told me. "So we did the experiment. We thought, We'll see how experience guides language learning. But what happened was that the blind children learned things they shouldn't have been able to. They knew the answers to things beyond their ability to experience. That was very upsetting. Well, we were happy at this victory of the human spirit but unhappy at having wasted our time with blind children. I figured the experiment had failed—simple as that! I went to my husband, Henry"— Henry Gleitman was then the chairman of Penn's psychology department—"and he said, 'So how *did* the kid learn the answer?' I said, 'Oh, that's not important,' and I went to Cambridge to talk with Chomsky. He was very interested. He said, 'So how *did* the kid learn the answer?' This was a little epiphany to me. I said, 'Oh, boy, I'm in trouble. Chomsky the mad rationalist and Henry Gleitman the mad empiricist agree on this.' So we went back, and the only explanation we could find was that the child was being guided by syntactic rules within the question—rules he already understood. The syntax tells the answer."

To the linguists assembled in the Stanford auditorium Gleitman had said,

> I've done everything I could think of to kids to show that they were responding to the world, and not to some inner quality. We started testing the effects of good and bad mothers, but they didn't have any effect. So we ripped the ears off of kids—we tested deaf kids. Then we tore their eyes out. Still, you know what? The little bastards learned language. The human child has a massive resistance to conditions, because he is going to learn language no matter what. You take away language, he invents one. We even did a nice study of preemies. They have the same experience in the world as full-term children do, but they're at a different physiological stage. It turns out that the age since conception is better as an indicator of language performance than the age since birth. Now, surely, observation of the world is one source of evidence. You can't take all forms of perception away from children. If you did, they would be falling off ledges and mistaking tigers for kitty cats, and pretty soon there wouldn't be any more children. But children aren't learning language from experience. They learn *words* from experience. They bring the sentence with them.

In the innatism to which Gleitman was a convert, the Three Days question was not "How do children learn language?" but "How does language flower out of the child?" What happens in the mind to permit that burgeoning comprehension? Gleitman had already found a piece of the puzzle: she showed that the Three Day clock is set at conception. But when does the clock run down? Is there a set deadline to language learning? This was the question to which Genie's arrival was so exquisitely timed. It burst into prominence in 1967, three years before her discovery, with the publication of a book by the Harvard neuropsychologist Eric Lenneberg called *Biological Foundations of Language*. The book was in some ways more revolutionary than Chomsky's of a decade earlier—more revolutionary for being more concrete. Lenneberg played Lenin to Chomsky's Marx, Itard to Chomsky's Condillac.

"Lenneberg's book was beautiful, marvelous," Catherine Snow once told me. "Chomsky's brain, the linguist's brain, has no nerves in it; Lenneberg gave it a biologist's brain, with a cortex and lobes and axons and dendrites."

Chapter 4 of *Biological Foundations of Language* presented what has since been called the critical-period hypothesis. It suggested that the brain is able to learn a primary language during a certain early period, and not later on, and it proposed physiological explanations of why this might be so. Lenneberg's innovation lay in those explanations; the idea itself had been around for a while.

"Lorenz proposed it," Jay Shurley told me. "Konrad Lorenz was an ethologist who discovered that he could train baby ducks to follow him around if he trained them at a certain period. That was ducks. In humans, it is almost impious to raise the question of a critical period. The creationists would have you not raise it. But it's been raised. Piaget did his lifelong study about what ages children develop certain capacities. The theory's as old as St. Augustine, who realized it in an intuitive way back in 600 A.D. when he said, 'Give me a child until he is six, and I'll give you a Catholic for life.' Augustine was wrong," Shurley said. "It takes till twelve."

Lenneberg agreed: It takes till twelve. According to Lenneberg, the child's ability to learn its mother tongue effectively ends at the onset of sexuality. Chomsky had not concerned himself with critical periods. Nevertheless, if Chapter 4 were to be borne out, it would have the effect of vindicating Chomsky, for how could language be tied to our biological clock if it weren't tied to our biology?

His concreteness notwithstanding, Lenneberg was, like Chomsky, a theoretician. What was needed was a clinician's validation, but the clinician would need something to work with: a child who had exceeded Lenneberg's deadline—who had passed twelve and hit puberty—but was still embarking on learning language for the first time. After 1967, there was a yearning in the linguistic field for a proper young arbiter— someone who could do for Lenneberg and Chomsky what Victor of Aveyron had been meant to do for Condillac.

16

The accounts in Susan Curtiss's dissertation of Genie's progress in the hospital during the spring of 1971 are all secondhand, gleaned from videotapes and interviews. Until after the consultants' conference in May, the UCLA graduate student and the subject who would shape her career had not even met. On

June 4, 1971, that situation changed: Curtiss accompanied Victoria Fromkin on a visit to the hospital.

She found the setting itself daunting. "I was never a person who thought of being a nurse or doctor," she told me. "I've never been comfortable in the children's ward of a hospital. I'm not good in hospitals. It's not my strong suit. I was also scared—or, at any rate, nervous." And with reason. To an unacclimated sensibility, Genie was a true grotesque. She was barefoot on the morning Curtiss met her, her tininess exaggerated by a dress that was too long, her movements jerky, her teeth jagged and discolored, her hair thin. Curtiss describes her as "pitiful and strange," and something else: pretty. The scientist was enthralled by the softness of the child's manner, her beautiful skin, the blush in her cheeks—"almost as if an artist had painted each one of them carefully and delicately"—and her upturned nose—"finely drawn like that of a china doll." She soon learned that Genie's indiscriminate spitting, scratching, nose-blowing, food-filching behavior could be somewhat less appealing. "It was hard," Curtiss said of the early contacts. "She was very—She was—hmmm—challenging."

The timing of Curtiss's arrival made her mission doubly difficult. Genie had not yet been trained into social acceptability, but in other ways she had progressed unfortunately far from her innocence of the autumn before. "In terms of watching Genie learn language," Curtiss said, "I felt I was arriving a little late."

Her tardiness was relative. If Curtiss had been at the hospital's admissions desk on the day Genie arrived, she would have encountered a languaged person, in the sense that all children have some degree of language before they begin making use of it. Genie could not have acquired her meager store of words if she had not previously mastered one of the most profound early tasks of any language learner: she had learned to separate meaningful sounds from the general cacophony surrounding her. In the words of Lila Gleitman in her address to the Stanford conference, Genie had "bootstrapped."

"The child has no passwords," Gleitman said on that occasion. "He doesn't know he's in the U.S. He doesn't know he's learning English. His mom shows him this room"—she waved a hand out over the audience—"and describes it. What does she say? 'Bahbahbahbahbahbahbahbahbah.' That's what she says. She could have said that the lady in back there is wearing blue, but what she really said, as far as the child knows, is 'Bahbahbahbahbahbahbahbah.' The question is: How does he figure out what his mother is saying about the room? OK? That's the story. That's bootstrapping."

What Gleitman calls bootstrapping is called other things by other linguists, depending on their academic orientation. But the mystery is the same: How does the child divide a stream of sound into syllables and sentences that he can begin to make sense of? It is easy to understand the child's bafflement. One has only to listen to an animated conversation in an unfamiliar language—our own language is built of discrete blocks, everyone else's of quicksilver. It seems as hard to grab a word out of a foreign tongue as to clutch a fistful of water from a pond. Yet the child, for whom all tongues are foreign, does just that.

Scientists are not yet sure whether the young listener first grabs phonemes—that is, individual speech sounds—or syllables, which can be made up of one or more phonemes. In normal conversation, nine hundred phonemes race by each minute, and there is attached to most of them no meaning to indicate their significance. Words have meaning, but their variations in length and form are countless, their boundaries indistinct. In normal speech, we break words up and slur adjacent words together; sometimes we pause within words. And if words are devious, sentences are even more so.

Here, as elsewhere, babies seem to know more than linguists can explain. Babies are born with some feeling for or understanding of language on both the phoneme and the sentence levels. Among the hundreds of phonemes used in the world's known languages, only forty are found in English. Newborns in English-speaking families display a preference for those forty, possibly from having heard them in the womb. They respond to their mother's native tongue. As the child ages, that discrimination becomes more pronounced; the child becomes more and more of a specialist. An adult speaker of English cannot accurately hear the phonemes peculiar to Chinese or French, much less replicate them in speech, without intensive training. It appears that the newborn doesn't so much develop his predilection for his mother tongue as let his perception of "foreign" phonemes atrophy. The Chinese baby is born with a developing bent for his native "r"-less language, but he can hear and pronounce "r." An American baby can do the same for all the French vowel sounds.

An equally astonishing ability applies to sentences. In the mid-1980s, Kathy Hirsh-Pasek, who studied at the knee of Lila Gleitman and now teaches at Temple University, was frustrated by one of the standard constraints of linguistics research: most testing is done verbally, and therefore only children who already have some verbal ability are tested. What, she asked, did the prelinguistic child know? She and two colleagues devised methods to measure the responses of very young subjects. They played tape recordings of sentences to nine-month-olds and observed eye movements for telltale indications of recognition. When the sentence ended at the proper place, the child acknowledged it. When the sentence ended improperly, the child did not recognize it as language. The incorrect sentence was received in the same way as arbitrary noise. Hirsh-Pasek has applied this method to younger and younger children. She professes surprise at the further results. Infants of four and a half months can tell correct from incorrect sentences, and what's more, they can do so for sentences in both Polish and English. The tests suggest that the ability that the nine-month-old has in its mother tongue the infant may have in any language. It has not yet let languish the grammars it will not use.

17

Though Genie had embarked on language learning before Curtiss met her, she hadn't acquired enough to make her available to the standardized tests that determine children's linguistic competence. In the summer of 1971, Curtiss and Fromkin faced the task of inventing a completely new set of linguistic examinations appropriate to Genie. They eventually devised twenty-six of them. The administration of those tests, along with a battery of psychological and neurological tests, would within five years make Genie, in David Rigler's words, "perhaps one of the most tested children in history."

Fortunately, the linguistic-research tradition allows for other, less rigid methods. Curtiss began a diary on the day she met Genie, recording everything that Genie said and analyzing it for signs of progress. Even here, Genie was stubbornly enigmatic. Most of the time, she said nothing; her vocalizations were usually whimpers or squeaks. "She had been beaten for vocalizing," Curtiss explained to me. "So when she spoke she was very tense, very breathy and soft. She couldn't be understood. There was a lot of sound distortion, as though she had cerebral palsy, but there was no evidence of muscle or nerve damage. Also, she had a high fundamental pitch. It was so high that we couldn't analyze it on the instruments we use to acoustically analyze human speech. And she was monotonic—high monotone. No pitch variation whatsoever."

Realizing how fruitless any attempt at formal research would be for the moment, Curtiss settled in for a summer of watching—getting to know the child and trying to gain her confidence. She sat with the patients in the rehabilitation center and, usually accompanied by Rigler or James Kent, took Genie on excursions.

"I would go by and take Genie for walks, or take her out to fast-food restaurants," Kent recalled. "At first, a nurse would go along with us. The nurse and I were supposed to be like surrogate parents, giving Genie the feeling of a family structure. We would hear some language from her on these trips, so Susan Curtiss started coming along to hear what Genie said. Genie was soon attached to Susie more than to the nurse who was supposed to be her surrogate mom."

The itineraries gradually expanded: they went to the zoo; they went for walks in Griffith Park. Especially, they went shopping—an activity Genie liked so much that on the way to the shopping center she would point to every passing building and repeat one of her new words, "Store?" The local Safeway and a Woolworth's were Genie's emporiums of choice, and there she displayed to Curtiss her disconcerting brilliance at both offensive and charming behavior. She would attach herself to strangers whom she found interesting, grabbing their arms, putting her face directly in front of theirs and staring into their eyes. Or she would attach herself with equal fervor to their possessions, from which Curtiss would have to pry her loose.

One piece of merchandise she found irresistible was beach pails. At her insistence, Kent had bought some pails she found especially attractive; she kept a collection of twenty-three of them, safely beside her bed at the rehab center. On an outing in mid-June, Kent used Genie's fascination to demonstrate a linguistic curiosity to Curtiss—a problem of definitions. He pointed to one plastic pail and asked Genie what it was. "Pail," she said. He pointed to another, and she said, "Bucket." There was no discernible difference between the two, but Genie was resolute in her distinction. The pails were located in a section of Woolworth's that Genie found especially enticing—an aisle of bright-colored plastic containers. Along with pails and buckets she coveted plastic necklaces, plastic purses, plastic trash cans—anything made of plastic.

When I asked David Rigler about the preference, the explanation upset him. "I think it was because of the bright colors and the texture," he said. "We learned that during her isolation Genie had had some small plastic toys. She had had a plastic raincoat hanging on the wall across from her potty seat." He paused, and then rushed on. "You visualize this house, and you picture this kid seated in this room, day after day, with very limited stimulation. She's grasping for some kind of stimulation, and the things she can see play a very large role. There's a plastic raincoat on the opposite wall." Rigler bowed his head suddenly and his shoulders shook, as though he were shrugging off something unbearable. "She liked plastic," he concluded.

For Genie, the excursions were visits to a magic kingdom. "I took her to walk at Griffith Park," said David Elkind, who in addition to attending the May conference had paid Genie two long visits. "Everything was so new to her. The brass on doorways, the animals, the birds. Some dogs barked and scared her. There were some students having a picnic on the grass and they saw her and saw that there was something special about her and they gave her an orange. She had never seen one before. She looked at it. She smelled it. We stopped in a store on the way home and she was entranced with the cellophane bags, crinkling them, listening to the sound they made."

Her innocent questing elicited extraordinary responses. A butcher at the Safeway saw how fascinated Genie was by the shrink-wrapped meat packages. He opened the service window and held out to her an unwrapped cut of steak, and she fondled, smelled, and studied it. In like fashion, over the months, he offered for her inspection bones, chickens, fish, and turkeys, all wordlessly, as though he and she shared a tacit understanding. Occasionally, when Curtiss reached the checkout counter the cashier would produce a toy or a trinket, with the explanation that "the man ahead of you sensed she wanted this and bought it for her." The gifts were chosen with such uncanny accuracy and were tendered in such silence that Curtiss became convinced that she was witnessing a preternatural communication—an explicit, unvoiced understanding—that her careful notebook analysis was unequipped to explain.

"Genie was the most powerful nonverbal communicator I've ever come across," Curtiss told me. "The most extreme example of this that comes to mind: Because of her obsession, she would notice and covet anything plastic that anyone had. One day, we were walking—I think we were in Hollywood. I would act like an idiot, sing operatically, to get her to release some of that tension she always had. We reached the corner of this very busy intersection, and the light turned red, and we stopped. Suddenly, I heard the sound—it's a sound you can't mistake—of a purse being spilled. A woman in a car that had stopped at the intersection was emptying her purse, and she got out of the car and ran over and gave it to Genie and then ran back to the car. A plastic purse. Genie hadn't said a word."

Genie's more conventional communication was improving. She still spoke in one-word snippets, but with an enhanced vocabulary. She was catching on to the give-and-take of conversation. She seemed, in fact, to have gained roughly the level that Victor had achieved at the Institut National des Sourds-Muets: she was forming social attachments and had picked up enough crude language (hers was spoken, whereas Victor's was written) to express her needs. Great attention had been paid all along, of course, to even the smallest signs of Genie's psychological state. When David Elkind met her, he noticed that she retrieved an item from her dresser drawer. "She had the idea of object permanence," he told me. "That's a major cognitive step for a child. Does something exist when it is not present to our senses? Children don't get that until after their first year." He also witnessed her attempts to bark like a dog she had heard earlier in the day. "That's a deferred imitation, and the delay is mediated by mental imagery," Elkind said. "So she was into her preoperational period."

"Preoperational period" is the terminology of Jean Piaget, the Swiss psychologist who believed that children have critical periods not just in language acquisition but in general mental development. The mind doesn't expand only by learning, he said. It unfolds naturally from within, going through predictable stages as the child matures. Preoperational thought is the second of those stages. Piaget saw the growth of language as tied to the growth of thought, as though it were a branch on the cognitive plant. Chomsky is inclined to see language learning and cognitive development as independent plants in a common garden. It was another dispute that Genie might shed light on eventually, but in the meantime Curtiss's evaluation of Genie's mental level concurred with the Piaget scale. The fervent search for names of things placed her at the beginning of preoperational thinking.

By all measurements, then, Genie was equipping herself to break out of her emotional isolation, her egocentrism. There might well be an intermediate step. According to L. S. Vygotsky, a contemporary and critic of Piaget's who applied the master's theories to language, the name-learning stage is followed by a period in which the child uses its new vocabulary to speak to itself, to encode its inner ideas. Vygotsky's theory embellished Heymann Steinthal's old formulation: perhaps, behind her inscrutability, Genie was building self-awareness—understanding herself as she was understood by others, for "that is the beginning of language." Through the summer and on into the fall, Susan Curtiss jotted down Genie's every utterance, all her sporadic, inchoate talk, and waited for the day when she might begin to reveal.

18

In July, the discontent over Genie's fate, that discontent whose fuse had smoldered and sputtered through the May conference, finally blew. The explosion would count among its eventual casualties everyone associated with Genie; it was set it off by a phantom case of German measles. Oddly, the most spectacular and determining event of Genie's first free summer gets no mention in Curtiss's dissertation. But it was documented by Jean Butler, Genie's teacher at the rehabilitation center, with whom Genie had developed a strong rapport. Butler's account was written in the form of a diary:

> *June 23, 1971*—I signed the necessary papers at the Hospital in order to be a volunteer and take Genie on field trips and to my home.

"Home" was a two-story, comfortably run-down house a block from the Wilshire Country Club, on Cahuenga Boulevard—a house that seemed neither modest nor exalted for a schoolteacher with an income of $13,000 a year. Jean Butler was doing all right. She had recently turned down an offer amounting to almost a quarter of a million dollars for twenty-five acres she owned near the Leisure World retirement village in Orange County. She was unmarried, and she supplemented her income by teaching summer school and, occasionally, by writing children's books. Her house had a guest bedroom downstairs, where Genie could sleep.

Not long after she had signed the papers, Butler called the hospital with dire news: she was ill, and her illness had been diagnosed as rubella. Genie had been exposed, and though she never came down with the disease she was at that point presumed to be contagious. Rubella is a havoc wreaker in schools, but in the light of Genie's past there was no humane way to isolate her. The obvious solution was to quarantine her with her teacher, and on July 7 she moved in.

"It was apparent that Genie was happy to be in my home," Butler wrote in her journal. But Butler herself was less than happy to entertain house calls from various members of what she termed the Genie Team. Butler's disparagement of Genie's other caretakers had been evident ever since the May conference at the hospital. She found Susan Curtiss inept, David Rigler self-important, James Kent overpermissive, and all of them ambitious and insensitive. For their part, the scientists saw Butler as needlessly contentious and personally troubled. Her running battles with the L.A. public school system, for which she worked, and with the doctors in the acute care section of Childrens Hospital, where she had been posted before her transfer to rehab, were renowned. She seemed as motivated as anyone by ambition. "Butler wanted to be the one who went into history with Genie," Hansen told me. "She wanted to be known as a miracle worker." In Curtiss's memory, the boast was explicit: she recalls Butler telling others in the rehab center that she, Jean Butler, was going to be the next Anne Sullivan, the young nurse famous for rescuing Helen Keller. Butler's protectiveness toward Genie struck Curtiss as proprietary.

July 8—Student Susan Curtiss was in my home recording speech and attempting to amuse Genie. However, she followed the child and hovered over her most of the day. She had a notebook handy and discussed Genie's speech and lack of it and her eating habits in a critical manner in front of her. . . . That evening Dr. Rigler phoned and I told him that the "help" he was giving me in the house was not helping me.

James Kent may have annoyed Butler the most. Among Genie's abiding enthusiasms was a fondness for the same behavior that had helped bring Victor's education to a crashing halt: masturbation. She was uninhibited by any concept of modesty, and was frequently an embarrassment in public. A persistent Butler claim, and one of her more sensational, was that Kent, unwilling to constrain a child whose life had been disfigured by constraint, encouraged Genie in her habit—an allegation that Kent has denied.

The care and feeding that Genie received in the hospital had spurred her development, and not just in behavior. Among other physical transformations, she began developing breasts. Signs of her sexual maturity were splendid news on one front—to directly test Eric Lenneberg's critical-period hypothesis, Curtiss and Fromkin would need to observe the language-learning attempts of someone past puberty. But it was a heartrending serendipity. David Rigler once showed me calendars he had made to follow Genie's progress in conquering her bed-wetting. They illustrated eloquently the child's awful dilemma. There amid the dry days and the wet days were marked the days she had her menses. She was getting her period and being toilet-trained, all at the same time.

"I expressed my fear to Dr. Kent that Genie was being experimented with too much and not being allowed to relax," Butler recounted in her journal. "He said this was necessary." Butler did not feel that she was alone in her concerns:

July 13—Sue Omansky of the Department of Public Social Services visited my home. . . . [She] was extremely critical of putting this child on display as a guinea pig and objected to the UCLA student hovering and jotting down everything said by the child. Miss Omansky expressed her belief that these men were using Genie to gain fame.

As the summer progressed, the tensions between Butler and the scientists sometimes erupted into full-volume arguments. Her house became the field for a jurisdictional battle of Titans. Sue Omansky, in her position with the Department of Public Social Services (DPSS), was Genie's de facto guardian. Her department had little inherent interest in making Genie accessible to researchers from Childrens Hospital; still, the two institutions were bound together in Genie's name. They had been conferring for months about how to get the child out of the rehabilitation center and into a private home. Now the rubella had forced the issue. Butler applied to the DPSS to become Genie's foster parent, and Omansky felt that the teacher's home was suitable for a permanent placement. But her DPSS supervisors, after their discussions with Childrens Hospital, had reservations. For one thing, it was against hospital policy to place patients in the homes of people who worked at the hospital. For another, it was felt that Genie would be better off in a home with a foster father as well as a foster mother.

Butler had a handy solution to that problem: she decided to ask her lover to move in. He was Floyd Ruch, a psychologist who had taught for thirty years at the University of Southern California and had written a seminal textbook, *Psychology and Life*. He was well-to-do and well thought of, but he was not unencumbered. Ruch was separated from his wife and was living alone, two blocks from Butler's house. In effect, though, he was already on the scene—enough so to be drawn into some of the quarreling between Butler and the Genie Team. Butler's journal recounts a disagreement between herself and David Rigler that turned into a late-night shouting match on the front walk, with Ruch rising to break it up, telling Rigler that no argument was settled after midnight and to go home. It's an incident that Rigler doesn't recall. "Oh, something like that might have happened," he told me. "We did argue about administrative stuff. But not shouting. And not at midnight."

July 14—I asked Dr. Kent to have Miss Curtiss removed from my home, as she was no help but completely untrained and inexperienced with children and had no awareness of safety factors. Dr. Kent said it was necessary to have her here and the need for phonetic recording of speech attempts was more important than her lack of ability in helping with

Genie. I pointed out that Genie did not talk around Miss Curtiss.

A few days after that entry, at the height of the conflict, came the episode of the puppy. Rigler relates it this way: "At one point, I visited Jean Butler's home and had a golden-retriever puppy with me, and Genie must have seen the puppy through the window, because according to Butler she got very upset. Now, this puppy was only ten or twelve weeks old. It was just a fur ball, and it wasn't up against the window, it was still in the yard, but Genie must have been scared of it."

Butler's version is more vivid:

July 20—Dr. Rigler phoned and said his wife had picked up a puppy and he would like to bring it over to show Genie. I asked him to wait a few days. He said he was anxious. I then said to please keep the dog in his car and let Genie peer through the window. . . . At about 8:00 P.M., Genie and I were folding sheets and the task was giving her great satisfaction. . . . Just then Dr. Rigler came. . . . He took her hand and led her to the front door, opened it, saying, "Come with me, Genie, I have something to show you." By this time Mrs. Rigler had taken the dog out of the car and placed it on the lawn. From the porch Genie saw the dog and ran back in the house, slamming the door violently. She got in my bed. . . . For a while she watched the dog through the front window. The Riglers left and Genie stayed in my bed for two hours, frequently getting up to go to the bathroom. She said, "No dog," and "Scared." She slept less than two hours that night. At 2:30 she came in to me and took my hand and led me to her bed. I sat by her for two hours while she repeated "Scared."

Genie's aversion to dogs was famous even before the incident with Rigler's puppy; Rigler himself had witnessed it during his earliest walks around town with Genie. After one canine confrontation, Rigler had commented to Butler that he had never seen such fear in any child. "The thing Genie would do when she saw a cat or dog, she would climb you like a pole," he told me. "Or she would desert you altogether. You'd look around and she'd be heading for the white line in the center of the road, because it was equidistant from the yards on both sides. And she was bright enough to know that a dog behind a fence was behind a fence, but a cat behind a fence was not behind a fence at all." Floyd Ruch, in particular, spent some time trying to get Genie over her alarm. He watched episodes of "Lassie" with her, and bought her a battery-operated toy dog that barked and wagged its tail. Only years later did he and Butler and the Riglers learn just how deep Genie's fear ran, and why.

Through July and into August, the haggling continued. Butler struggled to control the intrusions of scientists into her home and, at the same time, struggled to be numbered officially among them. She requested a pay raise of 38 percent over her standard teacher's salary, and she also asked to be acknowledged alongside the researchers in their scientific papers. Rigler offered to include her in the grant application with a position of "teacher-therapist-mother-caretaker-research associate." Then he recanted, telling Butler the suggestion was "a poor judgment on my part."

19

Genie seemed to be the only one growing more relaxed. Photographs of her taken at Butler's house show her animated, cheerful, composed, content. She sits on a hassock with one tanned, hospital-braceleted wrist cradled in her other hand, and looks up with such confidence, so completely self-aware, that it is hard to believe she is not a normal child. In a picture taken on the back porch, her ponytails have gone sodden from playing under the hose, and she tosses toward the camera a grin of unbridled delight. As was Victor before her, she was generally entranced with water. She even went to the beach, where she learned to sample, at least to ankle depth, the terrifying enticements of the Pacific Ocean.

Butler reviewed Genie's progress that summer in her diary: she claimed that Genie was wetting the bed less often, with thirty dry nights out of thirty-seven, and that her masturbation had declined as she gained interest in other activities. Along with everything else, Butler wrote, Genie was talking: "The quality of her speech improved and the quantity increased at least tenfold. . . . I was able to get Genie to say 'Yes' appropriately. This she had never done before. Also, I was able to get Genie to verbalize when she was angry, by saying the word 'angry' and making a hitting motion in the air or hitting certain inanimate objects (such as a large plastic inflatable clown). This was her first verbalization of her hostilities and anger."

In a letter to Jay Shurley, who was now back at the University of Oklahoma wondering about the summer's events, Butler wrote:

> You asked me about Genie's speech here. The last two weeks Floyd called her "My little yakker." He often said, "You're going to grow up and be a yakker like Jeanie." She talked one evening for 45 minutes after a trip to the pet shop to get four fish. During the day we talked and even argued about ¼ of the time. She was using two- and three-word sentences. She used the negative appropriately, and when I told her that she would have to come inside if she did not stop putting water on the service porch she said "No come in.". . . . She often described an object with two adjectives . . . "one black kitty" . . . "four orange fish" . . . "bad orange fish—no eat—bad fish," the longest expressed thought. I'll tell you the saga of the fish and their demise when you are here.

Butler's self-congratulatory assessment of Genie's mental state was borne out by an evaluating committee from the NIMH. The committee noted a "striking improvement" in Genie since her transfer to Butler's home. "Rather dramatic behavioral changes have ensued," its evaluation stated. "A visit to the home by two site visitors substantially confirmed the positive behavioral patterns and adjustment within that setting." The visitors reported back to Bethesda that Butler's home "would be an excellent placement" for Genie. In the contentious milieu of Los Angeles, however, the verdict was less sure.

August 6—Dr. Rigler insisted on driving me home [from a meeting], which he did. On the way home, he said that I was not cooperating as a "trainee" and that he had never had difficulty with students before. I got very angry and told him that I certainly objected to being treated like a student, a trainee, and an idiot. I told him that it was not necessary to tell me why I was using certain methods of discipline with Genie. I explained that he had had the last eight months to handle her and had done a very poor job. I explained that the problems she presented were the product of his department and I think I could at least be respected as an experienced person.

August 9—Before the regular mail delivery I found in my mailbox a metered but unpostmarked envelope containing a ten-page letter from Dr. Rigler.

The letter, copies of which had been sent to Kent, Hansen, and Omansky, was a pained recapitulation of recent history—an effort to set straight what had been scrambled in all the acrimony. "Dear Jean, I am writing to express my concerns about the current situation," it began, and proceeded to defend the charter of the research from Butler's charges of exploitation: "This child is not for sale, but in our view and in the view of funding agencies, knowledge obtained from study of this unique child is important knowledge to be employed for humanitarian purposes." Rigler extolled the staff of the rehabilitation center, which he described as "one of the best institutions of its kind to be found anywhere," but he also endorsed Butler's claims as a potential foster mother: "In this regard, I would offer my opinion that Genie is receiving excellent and loving care within your home at the present time." Nevertheless, he bemoaned what he saw as Butler's lack of cooperation, and he discouraged her hopes of increased compensation: "It is not likely that any parent or foster parent of a difficult-to-care-for child is adequately compensated for the endless and extraordinary demands placed upon them.

"Of course," the letter stated, "if that personal [financial] circumstance makes it not possible for you to continue the role of foster parent, I will be deeply distressed because of its impact on Genie.

"I am hopeful we can achieve a good level of cooperation," Rigler said in his concluding paragraph, "but events of the past month leave me very dubious."

Four days later, on the morning of August 13, Sue Omansky and her supervisor from the DPSS arrived at Butler's house. They brought with them their department's final decision on her application to be Genie's foster parent. It had been rejected. Butler wrote in her journal:

> For about twenty minutes Genie knew something was wrong. She was very upset when I told her that she must go with Mr. Wodowski and Miss Omansky back to rehab. She said, "No, no, no!" I told her I loved her very much but she must do what I say and go with them.
>
> Just before Mr. Wodowski took out her clothes he thanked me for all that I've done for Genie. . . .
>
> They left at about 10:30.

No sooner had Genie been taken back to the rehabilitation center than she was turned over to her new foster parents. Apparently, the policy concerning patients' living with hospital employees was a flexible one: the foster parents were David and Marilyn Rigler.

The sudden end of Genie's short summer on Cahuenga Boulevard marked a determining moment of sorts for Jean Butler. Her defeat confirmed her in the struggle against Rigler and the other members of the Genie Team. She began a relentless campaign to avenge the wrong that she felt she had suffered, firing off to various scientists letters critical of the team's research, and muckraking through the grant proposals and symposium papers of team members for the least hint of misfeasance.

Her first move was to complain to the DPSS about the apparent reversal of its position, claiming that the caseworkers had forsaken their better judgment and capitulated to pressure from the scientists to place the girl in an environment less hostile to research. Rigler recalls that the charge got reprimands placed in the workers' files, but it had no effect on Genie's placement, and Rigler dismisses it as vitriol. He dismisses also Butler's insinuation that he played an active part in her rejection. "I was surprised when they turned Jean down," he told me. He was also worried. Marilyn remembers the long drive she and David took—winding out to the coast on Mulholland Drive— to talk over their concern: the DPSS, in turning Butler down, had suggested no worthy alternatives for Genie's placement.

Not surprisingly, there is little coincidence between Rigler's accounts of the summer's events and Butler's; the breach between their views would grow chasm-wide with time. "She was angry at being turned down," Rigler said one afternoon as he, his wife, and I sat in his kitchen. "She began accusing us of bizarre behavior, but we found *her* behavior bizarre. She was as destructive as she knew how. She became the Wicked Witch of the West from then on, as far as we were concerned.

"We never had any intention or plan to be Genie's foster parents," Rigler continued. "Howard Hansen had discussed the idea with me. My wife and I consulted our respective navels, and each other's navels, and retired to our individual corners to think this out. And we decided to take Genie if no one else could. We told the social services department that if they absolutely couldn't get anyone, we would take her in for a limited period of time, that being—oh, how long, Marilyn?" He turned to his wife.

"Oh, a year."

"No, no. It was much shorter. I think it was three months. And then Genie arrived. I remember the date—it was Friday, August the 13th. And she stayed with us for four years."

III

WHEN SINGING
WAS ALL FOR
HER BENEFIT

20

If Genie's journey could have been read as the progress of a Horatio Alger figure, her residential history alone would have qualified her for the pages of *Forbes*. In little over half a year she had risen from mendicancy through an extended stay in an elite institution to a house in Hollywood. Now she arrived at the nicest accommodations yet.

David and Marilyn Rigler lived in Laughlin Park, an exclusive enclave of sixty-one homes in the Los Feliz district of Los Angeles. Beginning in the silent film era, the neighborhood had been popular among movie people and other entertainers because of its proximity to the Hollywood studios. Cecil B. deMille built an estate there, where one of his daughters still lives. Charlie Chaplin, Anthony Quinn, Lily Pons, and, in more recent years, Andre Kostelanetz and Chick Corea have all called Laughlin Park home. Rigler liked to walk visitors down the street past the house where W.C. Fields once lived. The neighborhood has remained a self-conscious exception to its surroundings—self-conscious enough that in recent years a gate has been erected at each of its entrances. Within, the streets are secrets: hushed, crooked, and close. Manorial houses hide behind massive boxwood hedges and high stuccoed walls.

Laughlin Park is an island of prestige built on a hill of prominence. It looks out over the petty industry and striving congestion of the flatland like Mont-Saint-Michel over the murderous mudflats of Bretagne.

Over the decades, Laughlin Park has fallen in and out of vogue. During the time the Riglers lived there, it was in something of a slump, no longer fashionable among the movie set, and not yet as tony as it has since become. But it was still popular with doctors, who found it less expensive than Bel Air and convenient to Childrens and other hospitals. The Riglers' house (whose yard, to the rear, faced the back of Jack Dempsey's old home), was not especially glamorous for Laughlin Park, nor was it notably humble. One friend, recalling a kitchen under interminable renovation, kidded Rigler that he had the "Tobacco Road of the neighborhood," but other visitors described it as a professor's house: respectable in its outward appearance; comfortable, cultured, and genteelly cluttered within. At least until Genie arrived, it was an orderly sort of place. David and Marilyn had three adolescent children, a cat, and Tori, the golden-retriever puppy whom Genie had already met. Genie was given a downstairs bedroom and a bathroom of her own. There was a large backyard where she could play and even some neighbors she could visit: the Hansens also lived in Laughlin Park, several blocks away, in a house that once belonged to silent film star Antonio Moreno.

The presence of a new family member occasioned immediate adjustments for the Riglers. "For one thing, we prize books," Rigler told me. "Genie's room was a room in our house that had been a sort of library. Two walls were filled with books and magazines. Genie was fascinated by them, especially the *National Geographics*, and she had her favorite issues. She could also be destructive. I can't bring myself to mark passages in books. But if she liked a page she might just tear it out."

And she might just do other things as well. On her arrival at the house, Genie ran her fingers nervously around the perimeter of each room, then defecated in Rigler's daughter's wastebasket. She urinated every ten minutes, wherever she happened to be. That habit eased almost immediately, but others didn't. She hid feces in her room (she had also done this at the hospital—once, to Rigler's great amusement, spraying them with deodorant to mask the smell) and appropriated possessions of the family's other children. She sat at the table with her cheeks bulging, waiting for her saliva to break down the food that she had still not learned to chew; that worked passably well with the cereal and apple sauce she was accustomed to eating, but as Marilyn Rigler introduced tougher foods to her diet the method entailed copious spitting.

The Riglers spent the first several days trying to get Genie to accept her old nemesis, Tori. "We found that Genie and the puppy couldn't be in the house at the same time," David Rigler told me. "So we instituted a program where they could get to know each other. We had them on opposite sides of the sliding-glass porch door. Then, when Genie had got used to that, we opened the glass and left the screen closed, and then we opened the screen. She eventually reached out when the dog was turned the other way, and touched its tail, and from that time on she was fine."

The success of fur-ball therapy reinforced a general optimism. Genie was at last settled in a home; she was at last free of vituperative bureaucratic wrangling. The grant from the NIMH had come through. Over the next two years, it was to provide $100,000 through Childrens Hospital for a wide range of research efforts, including the language studies of Susan Curtiss and Victoria Fromkin. David Rigler, as the principal investigator, was released from his duties at Childrens Hospital for almost half his time, with no reduction in pay, to attend to his work with Genie. Under the grant's terms, his wife—who was working toward her graduate degree in human development—would be paid from $500 to $1,000 a month for her ministrations. Los Angeles County would also furnish the Riglers with foster home support amounting to $230 a month. From now on, the research could proceed unimpeded, the only constraint on its pace provided by Genie herself.

Susan Curtiss maintained at the Rigler house her almost daily visits, recording in her large red notebooks as much of Genie's speech as she could catch. When, at the beginning of September 1971, she began administering the first of a series of linguistic tests that she and Fromkin had devised, she found out quickly how exhaustingly stubborn and restless Genie could be. Even on the child's cooperative days, when she obeyed orders and participated in activities, she never initiated anything, and her participation was always minimal, always frustratingly efficient. She was, Curtiss decided, lazy. How was one to know whether such a child was really still at the one- and two-word-sentence level or was just disinclined to use sentences of greater complexity? Much later, when Genie began using sentences of several words, she would compress them into one or two syllables, so that "Monday Curtiss come" would end up sounding something like "Munkuh." This behavior earned her the nickname, among the linguists, of the Great Abbreviator. She would pronounce the uncondensed version only on firm request. Genie's capabilities, Curtiss decided, were "masked by her behavior."

Another masking behavior was so ingrained as to be metabolic. Genie was slow. Unless confronted with a dog or some other alarming apparition, she moved as though walking through water. This behavior had been observable from the beginning—ever since she shuffled into the social services office on the day of her discovery—but it became more starkly evident as her comprehension of verbal commands increased. When she was asked to do something, she would often not move at all until many minutes had passed, and then would suddenly obey, as though the request had just then registered. She had the same "latency of response" with language tasks. One day about three months after Genie arrived at the Rigler home, Curtiss said to her, "Tell Rita who you went to see this morning." Some ten minutes of further conversation transpired before Genie finally piped up with a name. There was no sure way to know whether the child could not answer a question or had simply not answered it yet.

Curtiss had taken to reading stories to Genie, to which Genie remained politely oblivious. Then, on October 13, the oblivion broke. Curtiss saw the girl's facial expressions reflecting the content of the tales. Genie had always heard; now she was listening. She was listening in general—tuning in to talk not aimed at her. In a word, she was learning to eavesdrop. As Curtiss and the Riglers became friends, Genie often seemed to be doing the observing while the scientists did the talking. Sometimes she would try to obstruct the conversations between the adults, but at other times she listened in and occasionally even interrupted with apropos comments.

Her new home was a fertile environment for such progress. In their parlor the Riglers had a Steinway concert grand. It was not often played by members of the household, but Curtiss, usually just before dinnertime, would give recitals for her audience of one. If Genie merely tolerated being read to, she was a rapt concertgoer. "Music sent her into a reverie," Curtiss told me. "She would be compelled to stand there, and may even have hallucinated. I don't know where she went. She may have been musing on the past." But Genie was transfixed only if the music was classical, and only if it was performed live. Rigler's explanation for this goes back to the years in the little room: during part of Genie's incarceration, a neighbor's child took piano lessons, and his practice sessions, filtering in through the barely opened window, were Genie's matinees. Whatever their source, Genie's tastes were adamant. If Curtiss's repertoire strayed too far into the popular, Genie would pull her hands from the keyboard and replace the sheet music with a piece she recognized as being more highbrow.

On November 10, Curtiss was playing and singing some nursery songs she had discovered that Genie would tolerate. To her surprise, Genie clapped, danced, and stamped her feet to the music when Curtiss asked her to, and she sang, changing pitch in a semblance of tonal control she had never previously demonstrated. A week later, music provided the context for another innovation—not in inflection this time but in volume. During a drive to the hospital, Curtiss sang Genie an improvised song about their destination. Genie joined in, repeating "hospital" over and over, and once, in defiance of her fear of vocalizing, belting the word out. Some months later, she defied that fear again, this time letting out a scream when David Rigler tried to remove some wax from her ear. The event went straight into the notebooks. As far as the researchers know, the scream was her first and her last. But coming from a child whose explosions were almost always underground, it was at least an honest scream.

Advances in speaking came packaged with behavioral leaps. The person unofficially in charge of teaching Genie how to behave was Marilyn Rigler. To show Genie how to chew, she chewed with Genie's hand held to her jaw. In four months, Genie had learned to move her own jaw in approximate fashion, and the Rigler dinner table recovered a semblance of normality, disrupted only by Genie's gesturing. Instead of asking for what she wanted, Genie would grab Marilyn's face or arm and then point or otherwise gesture to indicate her need. Her gestures were a kind of language, peculiar and peculiarly effective. To express pleasure, she would moisten two fingers in her mouth and rub them quickly against Marilyn's nose. But communication at dinnertime required conversation of a more conventional sort, and Genie was pressured into learning to state, not manually indicate, her desires.

After Genie had had a while to adjust to life at the Rigler home, she was enrolled in a nursery school and, later, in a public school for the mentally retarded. At home, she was given speech therapy and taught some sign language—in part because signing seemed to suit her predilection for manual expression. In general, though, she remained extremely taciturn. Curtiss and the Riglers saw no evidence of the chattiness or the long-string sentences that Butler had reported. Her lack of expressiveness was nowhere more dramatically demonstrated than in her tantrums, which she still conducted in a straitjacket of silent self-destruction. Marilyn Rigler painted Genie's fingernails, predicting, accurately, that vanity would discourage her from tearing at the walls and floor. Knowing how much Genie loved to be called pretty, she told her that she was *not* pretty when she scratched herself or ripped at her face. Marilyn found herself in the strange position, for a parent figure, of teaching a child how to have a good king-hell-buster of a fit—how to slam doors and stamp her feet. She dragged Genie out of the kitchen so that she could do her stamping outdoors.

Here, too, gesture gave way to word. In Genie's iconography, a shaking hand indicated frustration, while a single shaking finger signaled the imminence of a full-blown tantrum. Seeing these storm warnings, Marilyn would say to her, "You are upset, you are having a rough time." Soon she had only to say "You are upset" for Genie to assent, "rough time." Eventually, "rough time" became a verbal shaking finger, a spontaneous phrase by which Genie could broadcast distress. Curtiss witnessed a further breakthrough in emotional expression one morning when she arrived to find Genie crying. She had had a cough and a cold and had complained that her ear was aching, and she had just learned from Marilyn the scary news that she would have to go see a doctor. "I noticed the striking change in this girl who such a short time previously did not sob or shed tears," Curtiss wrote in her dissertation.

In mid-June 1972, Curtiss recorded an event that approximately marked the first anniversary of her acquaintance with Genie. As with other accounts in Curtiss's dissertation, it is hard to tell who, subject or scientist, was being more changed by the experiment. "Today I took Genie into the city," Curtiss wrote. "We browsed through shops for about an hour. We sang and marched and carried on in our own nutty, special way as we walked. Genie seemed elated and delighted by everything I did. She commented, 'Genie happy.' So was I. Our relationship had developed into something special."

21

Another anniversary came a couple of months later—that of Genie's arrival in her foster home. It was celebrated in a much more public way. The eightieth annual convention of the American Psychological Association (APA) was being held in Honolulu, and several of Genie's watchers flew there to attend a symposium chaired by David Rigler. In the Mynah Room of the Hilton Hawaiian Village, Howard Hansen delivered a paper about Genie's early life in Temple City, James Kent spoke of the eight months she had spent in the hospital, and Marilyn Rigler recounted the trials of the year just past, in an address she titled "Adventure: At Home with Genie." Then Victoria Fromkin related what she and Curtiss and Stephen Krashen, another of Fromkin's graduate students, had observed of Genie's language.

"By November of 1971, a year after she was admitted to the hospital, Genie's grammar resembled, in many respects, that of a normal eighteen- to twenty-month-old child," Fromkin said, and she delineated some ways in which that situation had changed. In the weeks before the convention, Genie had finally shown that she knew the difference between singular and plural nouns; when Curtiss said "balloons" to her, or "turtles" or "tails," Genie now responded to the final "s" and pointed to a picture of two balloons or turtles instead of a picture of one. Similarly, she knew the difference between positive and negative sentences. She understood the meaning of some prepositions, so that when Marilyn asked her where elephants are found she replied, "In zoo." She understood yes-or-no questions, and she used possessives of a sort; she could say "Curtiss chin" or "Marilyn bike." (Only after another half year did she figure out how to insert a verb and say, "Miss Fromkin have blue car.") Her comprehension and production had progressed from one-word- to two-word sentences, with an occasional three-worder thrown in. "Now, two-word utterances are very complex, when you think of what this entails," Fromkin told her Honolulu audience. "She wasn't just stringing together any two words randomly; the two words which she put together in her sentences were very strictly controlled and rule-governed. They were not random strings."

"Rule-governed" was code, a hint to the hip that Genie was in the process of pulling off a coup that would rock the linguistic world. The rough draft of Fromkin's speech betrays her expectations. "It is clear that Genie is acquiring the rules of English grammar," she wrote, and then amended that to read "some of the rules." On a later page, "Genie is acquiring syntactic rules" was penciled over to read, more firmly, "has acquired." And on another page came the declaration "Genie at this stage has a grammar." All three references were deleted by the time Fromkin reached Hawaii.

The possible significance of Genie's achievement was made

clear in another section deleted from the final speech: "This summary of Genie's syntactic and phonological development indicates that language acquisition can occur after the age of five and even after the onset of puberty. Genie's linguistic development thus seems to contradict the conclusions of some that language acquisition occurs during the period when cerebral dominance, or lateralization, is developing." Fromkin went on to mention the "some" by name. Genie was going to debunk Eric Lenneberg: she was going to learn syntax, even if the prevailing theory of the time said that she could not.

There was a certain personal justice in that. Lenneberg knew of Genie and professed no interest in studying her. He commented to colleagues that he felt the case was too muddy for good science, complicated by the emotional trauma of Genie's incarceration. Fromkin and Curtiss would have strongly disagreed with his argument. "At first, Genie's natural state was nontalking, and that state might have been a reflection of her emotional state," Curtiss told me, getting (as she tends to do on the subject) a bit emotional herself. "But as she grew socially, and acquired the ability to be happy and live life, it became clear that her problems with language were not related to any distress or emotion. I don't see how an emotional profile could allow some aspects of language to grow but not others. There are a variety of views of language acquisition. The one I can best tell you about is my own, though my view is shared by most generative linguists. That view is that emotion has little to do with it. Certainly Genie was an emotionally disturbed child, but that wasn't relevant to my concerns."

At best, Genie could have provided a flawed endorsement of Lenneberg's theory. But she was capable of a ringing refutation. If Genie could not learn language, her failure would be attributed ambiguously—either to the truth of the critical-period hypothesis or to her emotional problems. If Genie did learn language in spite of all that had happened to her, how much stronger the rebuttal!

And, for that brief time, learning language was what she appeared to be doing. In retrospect, the September 1972, conference in Hawaii seems the point at which the tide of optimism was taken at the flood. If François Truffaut had made *The Wild Child* about Genie instead of about Victor of Aveyron, this is where the story would have stopped and the credits begun to roll.

22

It can be said, in looking back, that the prospects for Genie's eventual triumph were already beclouded that summer of 1972. One portion of the orthodoxy of language acquisition is the notion that, no matter how slow or how fast children learn language, they all go through the same stages, in the same order. After children get two-word phrases, they are poised for an explosion. It is as though they had been pushing a sled up a hill, and all of a sudden they are over the edge and racing down the slope; their skills accelerate as abruptly as that. Genie had been using two-word strings even before her stay at Jean Butler's, but month after month passed and the explosion never came. She continued to plod along at a slow, sled-pushing pace.

I once asked Susan Curtiss how she felt as it dawned on her that Genie's language was not going to explode. "It didn't dismay me, because I had the expectation that Genie, because she was older, would be dealing with a different set of rules," Curtiss said. "It was not at all clear to me yet that she would be so limited."

One thing that normal children learn quickly is how to form a negative sentence. They begin by saying "No have toy," and proceed directly to the next stage, where they bury the negation within the sentence: "I not have toy." Then they figure out how to use a supporting verb and say, "I do not have a toy," and the prodigies contract the verb to "don't." Genie stayed stuck at the "No have toy" stage for almost three years, and four years after she was talking in strings she was still speaking in the abbreviated nongrammar of a telegram.

Nor could she ask a real question. Normal children are sometimes thought by their parents to be much too adept at what linguists call the WH interrogatives. But any child who says "Why?" at every turn is doing what Genie could not. Since February 1972, she had been able to understand all questions involving "where," "when," "who," "how," "why," or "what." But when she was pushed to produce such a question herself, she mouthed monsters: "Where is may I have a penny?" or "I where is graham cracker on top shelf?"

One of the obstacles to forming true questions lay close to the core of Chomskian theory. To make a WH question, one must engage in "movement"—that is, deriving the word order of the surface sentence ("When is the train coming?") from the word order of the declarative sentence underneath ("The train is coming [soon]."). Movement was a facility that Genie did not have.

She also had a problem with pronouns. Most were missing from her lexicon entirely. "I" was her favorite, and "you" and "me" were interchangeable. Here the grammar reflected Genie's egocentrism—the lack of a border between her person and her world. She never figured out who she was and who was somebody else. "Mama love you," Genie would say, pointing to herself.

"Genie was highly motivated to interact socially and to use language in that interaction," Curtiss told me. "She could be almost frantic about it. She would stare at people's mouths as they talked. She was very inventive, very sensitive to whether she was communicating or not. For instance, she would often try to describe what she had done in phys-ed class at school. It's hard to do. It's an area where tense markers are needed, and where you have to indicate who's doing what to whom. And an area where she couldn't make herself understood. She would draw pictures, mime, use homonyms—try anything to get you to understand. If you thought you did but it wasn't what she had in mind, she would try again. She was very intense about this."

That Genie's language seemed motivated by her social strivings contained a pathetic irony, because she was especially incompetent at the array of interactions known as automatic speech—the interactions essential to social discourse. She could not learn to say "Hello" in response to "Hello," could not grasp the meaning of "Thank you." She would come when she was called but, with rare exceptions, could not summon anyone herself. She complained of a boy who was pestering her in school, but no one was ever able to teach her how to ask him to cut it out. Apparently, the words she brought with her from her imprisonment in the little room, Stopit and Nomore, were words she could imagine being aimed only at herself and her actions, not words she could use to defend herself. In all, Genie inhabited a prison not unlike a stroke victim's, with more to say than she was able to say, and aware of her inability.

Nonverbally, however, she had no such handicap. "Without a word," Curtiss wrote, "she can make her desires, needs, or feelings known, even to strangers." Rigler witnessed replays of the benevolent stranger syndrome. A woman in an idling convertible handed Genie her faux pearl necklace. A father and son walked by them carrying a toy fire engine, and suddenly the boy was back again, his fire engine offered in outstretched hands. Rigler's eyes tear when he talks about it.

Over the course of the three years following the convention in Hawaii, Genie's hoped-for linguistic ascent never materialized. When I asked Curtiss at what point Genie leveled out, she said, "Almost immediately. But it took us several years to realize that."

Faced with Genie's failure, many scientists have fallen back on the explanation—put forward by her father—that she was retarded. Curtiss disagrees. She noted to me that on some of the tests she and Fromkin administered Genie scored higher than anyone had ever scored. "On spatial tests, Genie achieved a perfect adult score," she said. "She could imagine a figure with pieces missing, and she could look at something from one perspective and know how it would look from a different perspective. She could draw silhouettes. She could categorize. Some people have said that categorizing is the key to learning language—that grammar is just organizing things into smaller and smaller categories. Genie could organize, but she couldn't learn grammar. Whatever she brought to bear on categorizing wasn't what she had to bring to bear on grammar. I would give her complex hierarchical models to copy, and she could do it effortlessly and flawlessly. Genie could apprehend the most complex structure.

"One time, we asked her to copy a structure made of a set of sticks. The sticks were different colors, but we didn't think about that—we were interested in the structure's shape. When Genie re-created the structure from memory, she got not only the shape but all the colors correct—every last stick—even though that was not part of the task. She could do all these things that are supposed to be related to grammatical structures, but she couldn't get grammar."

Genie's specialty—her ability with the spatial and the concrete—was reflected in her talk. Most children concentrate their conversation on activities and relationships: what happened when, what So-and-So did to So-and-So. Genie concentrated instead on objects, meticulously describing and defining them by color and shape, number and size. A normal child would rarely utter among its early several-word phrases the ones that dominated Genie's speech: "big, rectangular pillow," "very, very, very dark-green box," "tooth hard," "big, huge fish in the ocean."

In the late 1970s, as Curtiss finished her dissertation, she subjected Genie to a broad range of psychological tests that measured cognitive skills other than language, and she compared the results with those from tests administered to Genie by other scientists from the beginning. "I found some interesting things," Curtiss recalled. "I found that for every year that

Genie had been out of isolation she had advanced a year in mental age. Given a chance to interact with her environment, she was growing. This is the strongest evidence that she was not mentally retarded. You never see a case of a mentally retarded child in which the mental age increases a year with every year. Also, with retarded kids the lexicon is very impoverished. I have worked with one group of retarded kids who get a case correct but the semantics wrong. They're not sure of gender or number. Genie was always correct on cognitive matters. She knew how many and of what kind.

"Besides," Curtiss said, "being with Genie wasn't like being with a retarded person. It was like being with a disturbed person. She was the most disturbed person I'd ever met. But the lights were on. There was somebody home."

23

At home with Genie in Laughlin Park, the Riglers, too, felt that they were dealing with an intelligence. "This was not a dumb kid—no way," David Rigler told me. "She had energy and personality and incredible curiosity. She most emphatically responded to approval and was dismayed by reprimand. She craved affection and she gave it. She had a wonderful sense of humor."

Around the house, Genie handled complex tasks: she ironed, and she sewed both by hand and with a sewing machine. And she drew. Her drawings seemed actually to be part of her lexicon—a compensatory, self-taught speech.

When Genie was failing to transmit some idea, she would grab pencil and paper and sketch what she could not describe. She sketched more than objects: she could depict her thoughts and desires. Curtiss remarked on her ability to convey with a few deft strokes on paper the gestalt of a situation—the juxtaposition of people or things central to one of her tales. Her perception of gestalts was uncanny. Her mind had no trouble seeing the organization behind a chaotic scene or perceiving a whole from scattered parts. It was on the gestalt tests that Genie scored higher than anyone in the literature. But her portrayal of her complex comprehension was better achieved through visual than verbal means.

Throughout her emergence, Genie grasped her everyday experiences by relating them to images in magazines and books. When fear of the Riglers' pets was her greatest concern, she clipped photographs of similar cats and dogs and collected them, as though they had the magical protective qualities of voodoo dolls. When she saw a helmeted diver at Sea World, she did not calm down until she had got Curtiss back to the house and shown her a picture of the selfsame monster in *National Geographic*. Curtiss's early conjecture was that Genie had been programmed by a childhood that was almost devoid of event or society and was dominated instead by visual experience—an experience as static as a postcard. For her, the vision frozen in *National Geographic* may have been fully as alive as the one that moved at Sea World. Later, when investigations of Genie's brain unveiled the utter dominance of her "spatial" right hemisphere over her "linguistic" left, a more mechanical cause suggested itself.

Genie's progress was withal too slow to really be called

steady, but progress she made, through some idiosyncratic landmarks. She learned to fantasize verbally and to manipulate, and in March 1974 she combined the two skills and learned to tell an outright lie. She came home from school one day with a story about how her teacher's demands had made her cry. It was a fictional event, calculated to gain sympathy from Marilyn. A year later, when Curtiss caught Genie with a pocketful of stones (the which she was forbidden to carry), Genie said her pockets were full of "material." It was not quite a whopper, not by modern political standards, but a good solid start nonetheless.

Near Christmas 1971, Genie and Curtiss were walking down a hospital hallway when a small boy came up and began shooting at them with a toy pistol. It scared Genie, and when she and Curtiss had escaped out of range, she repeated a condensation of Curtiss's mollifying phrases, "Little bad boy," and, "Bad gun." On an evening at the Rigler home two weeks later, Curtiss was playing the piano and heard Genie mumbling and asked her what she'd said. "Little bad boy," Genie repeated. "Bad gun." Curtiss was pleased; for the first time, Genie was using language to relate a past event.

The question posed itself immediately whether Genie would be able to put into words events that had happened before words were part of her world. Would she have any memories from that time? And how would they be encoded? The answer—part of it—came all too horribly. "Father hit big stick. Father is angry," Genie said one day. And on other occasions, "Father hit Genie big stick" and "Father take piece wood hit. Cry." The scientists were learning about that part of the child's life they had not known, and learning it, moreover, from the child. "We worked with her fear of her father," Rigler told me. "We kept assuring Genie that her father was dead and was not going to appear and punish her. We had a problem communicating to her the concept of death. She was always afraid that he would return. As she learned to talk more, a stock phrase became 'Father hit.' Hundreds of times. Thousands of times."

Typically, one of her worst revelations was wordless. One day she would not come when she was called, and Rigler found her in her room sitting before a magazine, paralyzed with fright. The magazine was open to a photograph of a wolf. Genie was too terrified to explain her weird behavior, so when the Riglers had the opportunity they questioned her mother. They recall Irene's explanation—that on the rare occasions when Clark had interacted with his daughter he had imitated a dog, barking and growling at her. Sometimes, Irene said, he would stand in the hallway outside her closed bedroom door and bark.

The psychologists and psychiatrists familiar with Genie's case remain haunted by this image, and I have asked several of them, "Why a dog?" The nearest thing to an explanation was offered by Jay Shurley, and the explanation he gave me began, "I don't know."

"All I can think is that it had to do with Clark's appointing himself his daughter's guardian," he said. "Remember, he was going to protect Genie from the world, and at the same time he was punishing her with his protection. And people are often guarded by their dogs." He shrugged. "So he became a dog."

24

Since the November day in 1970 when Genie and her mother walked into the Los Angeles County welfare office, Irene had been a ghost in her daughter's life. She had never, perhaps, been much more—a blind, sad, momentary presence from the world beyond the door. Surely Genie could have understood little of her mother's own whispered existence. After the two escaped from their home, things had become better, and worse. It was not by any means merely an escape for Irene. If that had been all she was after, she could have escaped alone. But she confronted her husband and abducted her hostage daughter. If she had not had her daughter to take—had not had the obligation of setting right that blight on her life worse even than the injustice of her own mistreatment—who knows, Irene might just have stayed home.

Irene's belated heroism paid harsh dividends in the short term. "Heck, the first rattle out of the box there were headlines in the L.A. papers, and she was yanked into court," Jay Shurley said. "Her husband committed suicide. That was the first week. And then she lost control of the child."

Dismissed by the court, Irene returned to the house on Golden West Avenue. She spent the next five years traveling around greater Los Angeles, haunting the fringes of her daughter's celebrity. She visited Genie's various new homes and was introduced to her new extended family. Among the first people she met was James Kent, when she interrupted his initial session with Genie at Childrens Hospital. He described their introduction in his speech at the Hawaii APA symposium. "In the course of [Genie's play with a puppet], her mother and brother entered the room. She ignored her brother's greeting, moved quickly to her mother, and, pushing her face within a few inches of her mother's, peered at her without expression for a moment, then returned to the puppet play. . . . As we first observed it, Genie seemed less interested in her mother than in many of the other hospital staff. She would comply with her mother's requests to sit on her lap, but she remained stiff and aloof, and was noted at least once to have an angry outburst of scratching and spitting as soon as she could escape. Genie's mother seemed not to be aware of this notable lack of warmth; on the contrary, she remarked once after such an episode that Genie seemed to 'like me today.'"

Irene took to visiting the hospital twice a week, and as the visits went on they improved. "Genie's mother became more spontaneous and appropriate with Genie," Kent reported, "and Genie, as her relationship deepened with others, became more responsive and relaxed with her mother. Indeed, she began to look forward to the mother's visits with obvious delight."

The change was no accident. Kent credits the efforts of Vrinda Knapp, the hospital's chief psychiatric social worker, who began visiting Irene at home. Knapp's counseling of Irene was part of an attempt by the scientists to keep mother and child together. "We considered it important for Genie to have regular and frequent contact with her mother," Kent told me. "This was her only real link to her past, and we felt that it should be maintained."

The first battle the scientists had had to fight in that regard was keeping Irene out of jail. When she and Clark were indicted on child abuse charges, Howard Hansen prevailed on an acquaintance of his, a lawyer named John Miner, to attend the preliminary hearing on behalf of Childrens Hospital and argue in Irene's defense. For Miner, it was a fight on familiar terrain: the law from which he defended Irene was one that he had helped to write. He had headed a Los Angeles County committee on the battered-child syndrome, which drafted the legislation that made child abuse a felony in California; for those efforts, the governor had called Miner the "foremost protector of children" in the state. Miner was trained as a psychologist as well as a lawyer, and had run the medical-legal division of the L.A. district attorney's office. As part of his duties he attended and, often, assisted in, autopsies; he had helped autopsy Marilyn Monroe, Robert Kennedy, and Sharon Tate. He was an intimate inner player in the Los Angeles legal scene; attorneys of a subsequent and less gentlemanly era describe him as a veteran of a time in L.A. when legal business was more often settled informally among a close group of like-minded and personally acquainted attorneys, judges, and politicians.

The medical-legal division also handled child abuse cases, which brought Miner in contact with the doctors at Childrens Hospital and, through them, into contact with Irene and Genie. He took on Irene's cause shortly after his resignation from the D.A.'s office and found the case fascinating—after

Irene's hearing, he purchased as a "macabre memento" the revolver Clark had used to kill himself. Miner's involvement with Genie persisted after the disposition of the case, and in April 1972, he filed an application with Juvenile Court to become her legal guardian. An internal memo in the DPSS noted his concern: "His interest is motivated by his desire to safeguard [Genie's] part of her father's estate," it said. Miner explained to the regional DPSS bureau director that it would not be customary to become the guardian of a child's estate without also becoming the guardian of the child.

The estate left by Clark was hardly sizable. In addition to the house on Golden West Avenue, it included about $20,000, of which a third went to Irene and a third to each of his children. The court considered two affidavits: one from Irene consenting to the guardianship and one from Genie's "attending physician." "In said doctor's opinion," another DPSS memo said, "John Miner . . . would be a suitable guardian of [Genie's] estate and person." The attending physician was Howard Hansen. On May 18, the guardianship was assigned and, in June, Genie's money was transferred from Temple City to a savings and loan bank in Beverly Hills. Miner became the person legally charged with dispensing that money and with protecting Genie's interests, ensuring that she was not, for instance, abused or exploited by researchers. Researchers, for instance, like those at Childrens Hospital who had commended him to Irene and vouched for him as guardian.

The convenience of it all did not at first seem dangerous. Letting a patient live with a doctor, a subject with a scientist, was, of course, somewhat unorthodox, but Genie's case was an unusual one. True, the men in control of Genie all knew each other, but at least they all knew each other to be reasonable and honorable men. And, moreover, the goals of research and therapy were seemingly in concert. Why, then, should the boundary between them be sharply defined?

The first blurring of that boundary may have occurred with John Miner's presence at Irene's hearing; the hospital was, in effect, participating in a criminal case involving the family of one of its patients. By the time the Genie Team made the decision to rehabilitate Irene, the line was hardly discernible. Vrinda Knapp was instructed to glean from her counseling sessions with Irene a history of the family, and to relay that information to the scientists for their use. Many of the details in Hansen's paper at the APA convention, and much of what later appeared in Curtiss's dissertation, had been revealed by Irene to her therapist.

David and Marilyn Rigler sometimes drove Genie to Temple City on weekends, and those trips, too, were opportunities for observation. The Riglers frequently filmed Genie in their own home, eating, talking, playing; they also took a camera along to Golden West Avenue and filmed her with her mother. David Rigler once showed me some of that film. Genie is at the kitchen sink, beside her mother. Irene is working at the sink, her hair permed, her face a plain face, worn less with age than with worry. Genie flutters about her with a limby coquettishness, checking the counters and the refrigerator, occasionally coming to rest, like a butterfly alighting precariously, at her mother's side. In a fluty, urging voice she asks for cereal, but her mother says no, cereal isn't for lunch—they have chicken for lunch. As the camera follows, she leads Genie to the stove and lifts the lid on a large pot, so that Genie can see the chicken, and for a moment they are caught with their faces too close to the camera, frozen in grainy black and white. They are smiling. The mother's smile seems a little tight, but the child's is cheerful. When Genie walks off to a corner of the kitchen, the camera pans after her, and you can see her awkward hobble. She asks for orange juice, and for cereal again, and her high voice is all but lost in the roar through the kitchen window of the traffic on Golden West Avenue.

25

Irene's house had been rearranged and redecorated since the days of Genie's incarceration. "It looked very nice," Rigler told me, but other visitors found it depressing. The potty chair, at least, had been taken out back and burned.

Although Irene had lived there for more than a decade before her escape, she was only now getting her first good look at her own home. In the summer of 1971, she had undergone an operation to remove her cataracts, and her failed eyesight was largely restored. The surgery had been arranged by Hansen, Knapp, and Miner; like her psychotherapy, it was provided free of charge. Howard Hansen drove out to the suburbs to visit Irene in the hospital on the eve of her operation, and she talked to him of Mamaw and Dadaw and her childhood in Oklahoma, how she had gone swimming in the summertime, and how she wished that Genie could have the chance to do the same. When Hansen returned to check on her after her surgery, Irene got her first glimpse of him: he was not the giant his deep voice had led her to expect. His blue eyes, she thought, were the prettiest she'd ever seen. Over the ensuing weeks the other voices in her life took on substance, one by one. Knapp looked all white to her, angel-like. Rigler was unremarkable; Miner, amusing and exceedingly dapper. (He is given to French cuffs, monograms, lacquered nails, and occasionally, in the courtroom, spats.) He struck Irene as being honest, "as honest as a man can be."

When Irene arrived at Childrens Hospital for her next regularly scheduled counseling appointment, she learned that Genie would not be around for a visit. It made Irene want to cry, and made her angry; surely, she thought, the doctors must have known she would be starving to see, for the first time, her teenaged daughter. Finally, on the next visit, Genie arrived in the company of the Riglers, carrying a shopping bag. Irene was alarmed at how thin she was. Shy of showing sentiment publicly, Irene pulled Genie off to an upstairs office where she could give her a hug in private.

"They did a very good thing for the mother at the hospital, getting her the eye operations," Shurley told me. "Irene was very happy with that." But anyone who expected gratitude was in for a disappointment. "Jim Kent, in particular, went to bat for doing things for Irene," Shurley said. "I suppose Dr. Hansen did as well. Both were interested in converting her into a friend, but they didn't succeed."

It would have been a friendship across a great gap, as difficult to bridge as the chasm between Temple City and Laughlin Park. "Irene was quite looked down on, as the upper class can do toward the lower class," Shurley said. "It was a whole day's journey on public transportation for Irene to get back and forth from Childrens Hospital. She felt bad that she didn't have the right clothing—didn't have a dress to wear to visit her daughter in the hospital. Irene commented to me about this fancy hospital that her daughter was in—how she could not have afforded it if she had had to foot the bill. Neither side had an appreciation of what life was like for the other. Irene was suspicious of the Riglers' intellectualism. And I never felt that Rigler, for his part, saw Irene as human, saw Clark as human. Rigler, Hansen, Kent—they came from environments where they had always lived well. For them, Irene was like something the cat dragged in, and that was a problem for them."

Others who know the scientists protest that whatever social advantages Kent, Rigler, and Curtiss enjoyed had not been inherited. Curtiss was born to what she calls "the only impoverished Jewish family in Cleveland, Ohio." Coincidentally, a generation earlier, Marilyn Rigler had grown up in Cleveland also; after arriving there from the Ukraine, Marilyn's father had worked in the garment industry, as a pocket-maker, to save up to own a small corner grocery. David Rigler was, like his wife, first generation; his family came from Romania with nearly nothing. David's father, also, became a shopkeeper; he had a store on Manhattan's Spring Street until the advent of the Depression demoted him back to itinerant salesman. David was the first of his family to attend college—he began his undergraduate career at the age of twenty-five, on the GI Bill, after spending the years of World War II installing and repairing electronic equipment and ships' radars in the Panama Canal Zone. At first, he thought he would like to study engineering.

The Riglers are not among those scientists who confess an aversion to Irene, but those who do attribute it to something other than class difference: their lingering distaste for the role, however passive, she had played in her daughter's long imprisonment. From whatever source, Irene gleaned the impression that she was unacceptable by the standards of her daughter's new society. She was invited into the Rigler home a scant three times in four years, and even then felt her hosts to be nervous and abrupt. Her weekly visits with her child were conducted in some neutral place, usually a park or fast-food restaurant. Ironically, that arrangement was partly inspired by Marilyn Rigler's concern for Irene's feelings. Marilyn worried that Irene would be uncomfortable seeing her daughter in the home where Marilyn exercised a mother's authority and bestowed a mother's love. In the park, Marilyn hoped, the contrast between the adoptive and natural parents would be less stark, their standing a bit more equal. Afraid that it might be interpreted as condescension if she called Irene by her first name, Marilyn was careful to always refer to her by her family name; Irene, however, found the "Mrs." honorific formal.

In the unacknowledged class war, the person with diplomatic immunity was, strangely, the one most accustomed to a semblance of wealth: Jean Butler. She had grown up in a well-to-do Midwestern family and had traveled extensively in Europe. She and Floyd had married—she was Jean Butler Ruch now—and the couple had several homes and a yacht. "Nevertheless, I think Jean was more sensitive to that socioeconomic stuff than Rigler was," Shurley said. "She knew how to keep her distance, respectfully, and she didn't use her wealth and position to dominate the situation. She gave Irene advice, didn't usurp, didn't invade."

In contrast to the Riglers' respectful propriety, Jean Ruch liked to call Irene and gossip. Irene tossed off much of Ruch's talk as inconsequential chatter, but she found it friendly, and appreciated it. As Irene's health improved and she became accustomed to her life as a widow, her affection for Jean grew, and so, apparently, did her distrust of the scientists who were studying her daughter. One day, after her eye operation, she was leaving the rehabilitation center with Genie and David Rigler. They were walking slowly, to accommodate Genie's characteristic shuffle, and, as Rigler recalls, "We got outside, and Irene looked at her daughter and looked at me and asked me, 'What have you done to her that she walks this way?'" Rigler was taken aback. "I don't think Genie's mother ever understood what her role in Genie's condition was," he told me, and he noted that this denial may have been a testament to the success of Irene's therapy. "I think the mother, after her counseling and rehabilitation, had a task of her own—to resolve this in her mind in a way that would allow her to live with it," he said. "Irene saw our presence as a reprimand, an indictment—as a reminder. And we were too busy congratulating ourselves on our benevolence to notice how much we were antagonizing her."

26

As 1972 became 1973, and 1973 turned into 1974, David Rigler must have been well pleased with Susan Curtiss's progress toward her doctorate. Except for the linguistic work pursued by her and Victoria Fromkin, precious little was coming out of the ambitious experiment of which he was the principal investigator. During her years as a resident in the Riglers' house, Genie had gone from being "the most promising case study of the twentieth century" to being, in Rigler's words, "perhaps one of the most tested children in history." She had not, however, turned into much of an oracle.

"At one point," Rigler told me, "I did a diagram of all the people from around the nation who were involved with researching and helping Genie, and it was a huge circle," and he spread his arms as wide as they would go. The researchers had produced reams of data. But the data piled up uncollated and unprocessed, the sheer volume an impediment to the drawing of any significant conclusions. A handful of papers had ensued, most of them recapitulations of Genie's horrific childhood, and none of them of much more abiding import than the paper David and Marilyn Rigler submitted to the Twentieth International Congress of Psychology, in Tokyo in August 1972. The paper was titled "Attenuation of Severe Phobia in a Historic Case of Extreme Psychosocial Deprivation." It detailed how, by the use of such devices as a sliding glass door, Genie had been introduced to Tori.

The NIMH found the lack of progress troubling. In a series of site visits, its grant overseers expressed their concerns to Rigler. Worried that the data were being collected in haphazard fashion, they suggested new tests to fill in gaps, and asked that others be readministered. In the fall of 1973, Rigler was given a year's extension and additional money for "developing an adequate research plan" and analyzing the research he had already done. A year later, with the extension running out, the NIMH deliberated on his application for a further $226,000 to support the research for three more years.

Genie's progress was also being watched, from a greater remove and with a much more jaundiced eye, by Jean Butler Ruch, who gleaned reports of Genie's health and behavior from any available source. Convinced that Genie was not doing as well as advertised, she lobbied aggressively against Rigler, Hansen, and Curtiss with anyone in the scientific community who would listen.

Why did Rigler contend that Genie was acting appropriately in social situations, when she clearly was not, Ruch asked in her letter campaign. Why was Marilyn claiming credit for training Genie to set the table (by rewarding her with ten pennies each time), when Genie had already been a zealous table setter during her summer with Ruch, and before? Why, Ruch asked, did the Riglers say that Genie had arrived at their house unable to dress or clean herself, when the nurses had trained her to do all that at the rehabilitation center? Why were Rigler and Curtiss crowing that Genie was making three-word utterances by the end of her third year in Laughlin Park, when in the summer of 1971 she had been able to say "Foy big black car go ride" when she wanted Floyd Ruch to take her out to, for instance, the pet store, and "Bad orange fish—no eat—bad fish" in explaining why she had tossed her new pet goldfish out into the yard? Jean Ruch insisted that the Riglers had reset the chronology of Genie's progress to conceal the fact that Genie had declined in their care. "This sounds terribly self-serving," she wrote to one scientist, "but no one who saw her after her stay with us reports her ever as vibrant and active or acting and looking so 'near normal' as she was in our home."

Ruch charged that Rigler had inflated his original grant application with "imaginary consultants"—listing as collaborators eminent scientists who had done little more than poke their heads in while passing through. When I spoke to Rigler about this particular charge, he frankly admitted that he could not recall meeting one of the psychologists he had listed in the grant application as having spent two days with Genie; however, the listing of all these consultants could just as easily be ascribed to optimistic self-deception as to fraud.

Ruch also accused Rigler of callous behavior toward Irene; as evidence, she pointed to the Riglers' insistence that Irene visit her daughter in places other than the Rigler home. She complained that he had refused to abet those meetings with any financial assistance, even though Irene was running through her inheritance, had mortgaged her home, and was sewing and selling dolls to make ends meet. "Considering that Rigler et al. went all over the USA, Hawaii, and Japan on Genie Funds, to not give a portion of their State foster-care food allotment to the mother was [viewed as] unforgivable by all who knew her financial problems," Ruch wrote. In her files she catalogued this particular item under the heading "Mother's Need vs. Rigler's Greed." The files were voluminous, running, by Ruch's count, to six thousand document pages. "She used the Freedom of Information Act to go to NIMH and get all the records of my research," Rigler told me. "And then she got furious when they notified me that she had been given the documents."

Through the error of an inexperienced clerk, Ruch was sent a seven-page paper that should not have been released to her—the grants committee's appraisal of Rigler's application for a new three-year grant. "The rule is that under the Freedom of Information Act you may buy only documents about projects which have been approved," Ruch gloated to one scientist. She characterized the committee's appraisal as "scathing."

The NIMH grants committee met to decide on its recommendations in September 1974. A two-day site visit to Los Angeles had convinced the committee that "very little progress has been made" and that "the research goals projected probably will not be realized." Its report continued:

> The Committee feels that the proposed research plan is deficient in its own right and inappropriate for the special needs and circumstances of this unique case study. . . . The failure during the past year to implement the recommendations made by the Committee for which funds were made available . . . is disquieting. The Committee feels that this application is clearly lacking in scientific merit, and, therefore, unanimously recommends disapproval, requesting that its comments be conveyed to Dr. Rigler.

On the bright side, the committee expressed its opinion that the research had posed "no substantial risks to the individual who is the object of this proposal" and observed that "the therapeutic benefits to the subject have been and continue to be considerable." The well-being of the "subject" was nonetheless a worry:

> The Committee is concerned about Genie's future welfare and how the consequences of disapproval will directly affect Genie. The Riglers have indicated that without support for their research project, they would probably have to terminate their foster relationship with Genie and leave her future

care to the State of California. The Committee appreciates that Genie is properly a ward of California, not of N.I.M.H., and feels that the appropriation of research funds for Genie's maintenance outside of a research context would not be in her best interest or that of the Federal Government.

"There were some good reasons and some bad reasons for rejecting the grant," David Rigler told me. "But, essentially, they didn't understand. The study wasn't like most scientific studies. There were no controls. It's a study of a single case, and those are rare. They're anecdotal. They can't be done in the way of normal science." Early on, Rigler had scoured the literature for a good model of a long-term study of an atypical subject, and found only one of use: "This is a problem for which there is no agreed upon research paradigm, unless it be Itard's report of the Wild Boy of Aveyron," he wrote in a 1973 letter defending his work from his NIMH critics. Rigler didn't think that such single-case research could be conducted in the usual statistical format.

"The people on the NIMH staff are involved with grants," he told me. "I used to work with them, and I know what that means. There was pressure on me to be much more scientific in my approach. Measurements, that's what they wanted. Not that I didn't want to make measurements, but I didn't want to do so in ways that would be intrusive to the well-being of the kid. I was never able to satisfy people on the committee that I was doing this in the best way for science and for the child."

IV

LOST

27

On June 4, 1975, David Rigler addressed a letter to an administrator at Childrens Hospital summarizing Genie's progress over the past four and a half years. She was capable of some autonomy, he said, but she still needed substantial supervision. She could care for her hygiene and even prepare simple meals. Her self-destructive tantrums were less frequent. Rigler described Genie's performance on "a very large number of standardized and custom-designed tests, many of them [administered] repetitively over time," and added that, "the tests notwithstanding, Genie remains in some sense an enigma." She was still an emotionally disturbed child, he said, but there was hope. "At age 18, Genie has not stopped her process of achievement in any sphere," Rigler wrote, noting that she had "clearly established powerful emotional ties to both the foster mother and to her biological mother." He concluded, "As you know, we are contemplating relinquishing Genie's foster care; however, we have a continuing wish to be of service to her in a new placement."

Before the month was out, Genie's bags were packed. She went home to Irene—to the house on Golden West Avenue in Temple City, where she had spent the bulk of a painful childhood and almost every weekend of the previous six months. It was destined to be a short homecoming, for despite all the preparatory visits, Irene found the experience of living with her daughter difficult. Genie's masturbation disturbed her, as did the more normal teenage demonstrations of ignored requests and slammed doors. Irene imagined every irritating habit to be one her daughter had picked up at the Riglers. She soon felt exhausted and overwhelmed.

The Riglers had feared that she might. "After we gave her up, we were worried how the mother would take care of her," Rigler told me. "We have some money. We can afford babysitters and help. Irene was impoverished. So that first summer we made arrangements for Genie to go to summer school and, when that was over, to day camp. But the mother asked her, 'Do you want to go to day camp?' and Genie said no. So she didn't go. She stayed home, and before long the mother was calling for help. Not to us, but to the protective services."

Irene contacted the East Los Angeles Regional Center for the Developmentally Disabled, a division of the California Department of Health that deals primarily with retarded adults. The agency took on the case and began searching for a foster home to take Genie. The home they found was in nearby Monterey Park, and on November 7, Genie moved in. On the surface, the placement looked ideal. Visitors remarked on the new home's immaculateness, and found that Genie herself was kept clean and well clothed. There were two other foster children already residing there. But Genie's arrival brought immediate omens of trouble. She came bearing what one social worker described as "an amazingly large collection of plastic wastebaskets, refrigerator containers, and other plastic objects." Once Genie was at the foster home, the worker continued, "these 'toys' were put in the closet and she has not played with them since."

The family's cleanliness was accompanied by an almost militaristic rigidity. Genie was allowed little leeway in expression or autonomy. Irene's visits were treated as disruptions and effectively prohibited. Genie's reaction to the regime was to regress, seemingly intentionally, shedding by degrees the skills in comportment and communication that she had developed over the previous several years. She was especially distraught about Irene's absence. "Genie had a very strong connection with her mother," Marilyn Rigler told me. "She didn't remember her mother as one of her childhood abusers. She remembered her from those years as the person who kept her alive." Now, like her father before her, Genie had had her mother taken away, without explanation. And like her father, she retaliated by becoming a recluse—in this instance, however, she had only herself to lock away. She closed up, depriving the world of whatever she thought it wanted.

A barometer of Genie's happiness had always been her bathroom habits. Her lifelong bowel problems had waned at Jean Butler's house and returned when she moved to the Riglers', only to improve again as she settled in. Now they resumed, forcefully, and the consequence showed just how full circle her life had come. During her childhood, a chronic constipation had been Genie's physical protest. At one point, Clark had tried to remedy his daughter's obstinacy by forcing her to down an entire bottle of castor oil. The overdose had landed her in a physician's office. That battle, as it turns out, was premonitory.

According to Rigler, "the lady running one of the foster homes was rather bizarre." He recalled visiting the home "from time to time" and counseling Genie in her occasional outpatient visits to Childrens Hospital. "The woman was very rigid, and Genie had a powerfully strong will," he said. "Ultimately, the collision occurred over the issue of her toilet behavior. What happened in this home was that she became constipated, and this got to the point where it was very painful. The woman tried to extract fecal material with an ice-cream stick. There was no injury. But she was traumatized."

Genie's reaction to the trauma, as the scientists interpreted it, was to up the ante. If the world would go to that extreme to invade her sovereignty over her body, she would deprive it of something else—something it had desired from her and rewarded her for. For five months, she didn't speak. "Genie wanted to have some control over her life, and she never did," Curtiss told me. "She never had any control whatsoever over what happened to her. The only way for her to control her life was to withhold feces or withhold speech, and so she did.

"It wasn't an attempt to quit communicating that made her quit speaking. She had had this terrible—a couple of terrible experiences. She had a fear of vomiting, and she had vomited a couple of times and been punished for it. And then—oh, this story is so terrible I can't tell you all of it—she was in this foster home, and it was an abusive home, and they told her that if she vomited once more she would not ever get to see her mother again. She didn't know what she had done wrong, but she was afraid that if she opened her mouth she would vomit. But even during her elective mutism she wanted to communicate with certain people, and one of them was me, and, thank

God, she'd been taught some sign language. She signed furiously to me, about how much she loved her mother and missed her—about everything. You could see her wanting to eat, but she would refuse to open her mouth. It was very labored eating. She would—" Curtiss twisted her face sidewise and looked up, like a fish eyeing a morsel of food on the surface of the water. "And then she would open quickly and gulp it. After not eating, and living with that abusive foster family, she ended up in the hospital."

During the year and a half that Genie spent in her first foster home, Curtiss was the only frequent visitor from her previous life. Financial support for the linguistic research had collapsed with the NIMH grant refusal, but Curtiss still came by each Wednesday, observing her former subject, playing with her, taking her on walks, and teaching her new foster mother some rudimentary sign language. As the home scene steadily deteriorated, Curtiss became Genie's vocal advocate. The DPSS and regional center files chronicle the letters Curtiss sent and phone calls she made protesting Genie's placement and begging for help. "Over a long course I spent a lot of time meeting with people, talking on the phone, trying to get people to pay attention to the fact that Genie was in trouble," she told me. "I spent the better part of a year or more trying to get people to understand that she was in an abusive home."

Curtiss pleaded, also, with Rigler, and with Miner, who was still acting as Genie's guardian. Eventually, her warnings could not be ignored. John Miner saw firsthand how desperate the situation had become when he hosted a poolside barbecue at his house and persuaded Genie's foster parents to allow her to attend. It was the first glimpse he had had of his charge since her departure from Laughlin Park over a year and a half earlier. "The last time I saw Genie in the Riglers' home, my daughter and I had brought her a dress for her school graduation. She was very thankful. For the first time, she put her arms around me and kissed me on the cheek," Miner told me. "I'm sorry if I get a little emotional. It breaks me up even now. There was real hope there. Real hope that this youngster would have a life that would be worthwhile. Then, when she came to the barbecue, her behavior was just not recognizable. She would not speak, would not say a word. She snapped at her food as if a crocodile. As I recall, someone had to assist her to the bathroom. Her regression was just overwhelming. It was a shock to me. I had thought she would not flourish in the foster home setting, but I was not prepared for this. I wanted to reach out to her, to tell her there were people who loved her and cared for her."

Spurred by Curtiss, Rigler and Miner maneuvered to have Genie readmitted to Childrens Hospital at the end of April 1977. She stayed for two weeks, and her crisis eased somewhat. When she subsequently left the hospital for another foster home, her speech and eating had partially improved. But the intervention came too late to repair one piece of damage: to the esteem Susan Curtiss had held for Rigler and Miner. "Miner and Rigler didn't fight against me," she told me. "But they didn't help me, so it might as well have been the same thing." She saw their long inaction as an abandonment, and it mystified her, coming from people who had clearly cared so much for Genie. It mystifies her still.

"There's no adequate way to describe what the Riglers had gone through in taking Genie in," Curtiss told me. "They essentially sacrificed their family. Now that I'm a parent, I have a knowledge of how much they gave. Nothing was grudgingly given. They greeted her with open arms. The Riglers' home could not have been better. It was full of fun and enrichment, and my only regret is that Genie had to leave that environment for one so much inferior. Once Genie knew about the world of music, art, love, and culture, it was hard for her to go back to a meaner existence with a family that only knew how to sit around or do nothing. She suffered a lot. She was devastated at not seeing Marilyn Rigler. It was very clear that she needed to have those people in her life."

Curtiss's notes from Genie's tenure in foster homes transmit the girl's longing. "I want live back Marilyn house," Genie said in November 1975. In August 1977 it was, "Think about Mama love Genie." The notes were intended as records not of Genie's emotions but of her language ability, for Genie was once again the object of scientific scrutiny. In 1977, Curtiss and Fromkin received a grant from the National Science Foundation to continue their linguistic research. They were now the only scientists funded to work with Genie. As Curtiss told me, "None of the other research had panned out."

28

Curtiss's research was panning out on two fronts. While pursuing her testing—and even during the hiatus in funding, when she saw Genie largely as a friend—she was simultaneously compiling her doctoral dissertation, summing up the Rigler years, sorting out all the things that Genie had learned to do from all that she had not. "She had very quickly developed a vocabulary, and put her vocabulary in strings to express complex ideas," Curtiss told me. "She was a very communicative person. But, despite trying, she never mastered the rules of grammar, never could use the little pieces—the word endings, for instance. She had a clear semantic ability but could not learn syntax. There was a tremendous unevenness, or scatter, in what she was able to do."

That scatter had been one of the initial curiosities of Genie's case; now the years of research had seasoned it into significance. "One of the interesting findings is that Genie's linguistic system did not develop all of a piece," Curtiss told me. "So grammar could be seen as distinct from the nongrammatical aspects of language, and also from other mental faculties." Curtiss used Genie's exotic history like a chemist's centrifuge, spinning a complexity of elements out of the clear, simple liquid of childhood learning. Language emerged separate from other cognition, and fragmented within itself into a host of individual components. "In normal children, so much develops at the same time," Curtiss told me. "It's difficult to tell by observing the average child that acquiring language is a task separate from others, and full of discrete pieces. The hallmark of cognitive development in normal children is its multiplicity. Everything is going on at once. But we saw with Genie that these things could sprout independently, by means of different mechanisms."

When Curtiss says "mechanisms," she is not being abstract or metaphorical. She means not only psychological but physical mechanisms—structures in the brain. As Curtiss chased her quarry deeper into her dissertation, she chased it more and more in Eric Lenneberg's direction; her last chapter was on neurolinguistics, and it delved into the biological basis of Genie's language skills. Genie's inabilities bore out Lenneberg's theory, at least conditionally. She demonstrated that after puberty one could not learn language simply by being exposed to it. Her scatter was especially confirming. It divided the "learned" skills, such as vocabulary, from those said to be innate, such as syntax. Furthermore, the syntactic abilities, which both Chomsky and Lenneberg had predicted would be biologically determined, had indeed been constrained by Genie's biology—thwarted by her development.

It was a mischievous revelation. In affirming Chomsky, it might also be read as refuting him. As Catherine Snow of Harvard told me, "Genie's case could be taken as evidence for the empiricist position." And indeed, if some parts of language were innate and others were provided by the environment, why would Genie's childhood hell have deprived her of only the innate parts? How could a child who lacked language because she had been shut away from her mother be proof of the contention that our mothers don't teach us language? Why should she be unable to gain precisely the syntax that Chomsky said she was born with?

The problem was not peculiar to Genie's case. It was constitutional, an aspect of Chomskian thought that might seem, on its surface, paradoxical: if syntax is "innate," why must it be "acquired" at all? There was, of course, a constitutional answer: even innate characteristics must be developed. Though the shape of language is inherent to the human organism, the theory held, its advent might depend on certain agents and events, much the way the growth of beards, breasts, and body hair, though part of our species' design, awaits in the individual the summons of certain hormones. Taking that as a model, then, what summons language?

The answer might lie in Genie's brain. "It occurred to us that maybe Genie was not grasping grammar because she was using the wrong equipment," Curtiss says. As early as the fall of 1971, Curtiss, Fromkin, and Stephen Krashen had begun doing neurolinguistic tests in the hope of finding out exactly what part of Genie's brain they had been talking to all those months, what part of Genie's brain had been talking back. The equipment search would have alarmed those early linguists who thought that seeking a biological center for something as ineffable as language was as futile a misadventure as looking for a center of the soul. "It is useful to think of language as an organ of the mind," Chomsky had told me. But couldn't such a contention be an invitation to foolishness? Descartes, after all, had found it useful to think of the pineal gland as the connection, the clasp, between the body and its soul, and believed female beauty to be an excretion of the thyroid. How rational a rationalist does that make him sound today?

Nevertheless, modern neurology has found concrete mechanisms for other incorporeal things—or, at least, found where those mechanisms reside. The ability to watch a baseball's flight and know where it will land inhabits the brain's right parietal lobe, above and behind the ear. Getting a joke, understanding a metaphor, and realizing that something is inappropriate to say in a conversation are also talents of the right hemisphere. The right brain listens to music. Both hemispheres know the meanings of words. Mathematics, logic, and language—at least, the grammatical part of it—have a preference for the left hemisphere.

From the misfortunes of brain-damaged people, it is clear that language tasks are dispersed within their left-hemispheric home. Someone whose brain is injured above the left ear in a region called Wernicke's area may still be able to speak correctly, even glibly, but often there will be no discernible idea behind the voluble word strings. If the injury is forward of that, in Broca's area, the victim will struggle painfully toward expressing his thought, unable to form sentences or find words. Those words called function words will especially elude and frustrate him. "Function words are the ones there purely for grammatical reasons: no, if, and, or, but. Words like that," Helen Neville, a neuroscientist with the Salk Institute in La Jolla, California, explained to me. "Content words are nouns, verbs, adjectives, words that connote the content of a sentence. There is an enormous difference between the location of brain activity caused by content words and by function words."

Neville practices her neurology in a tiny fourth-floor lab so hidden away within the Salk Institute as to seem purposefully covert. The institute refers to the fourth floor as "interstitial space"; residents call it the "pipe room." It is a sort of mid-building basement, a dark wafer of windy gloom given over to the institute's custodial plant and to Neville. Children coming to her lab for testing follow a bread-crumb trail of wall-mounted cartoon characters through a forest of hissing air-conditioning ducts and moth-balled autoclaves. I have followed the same twisted trail, on several occasions, with the goal of learning from Neville some of the principles at work in the neurological study of language and watching, in progress, some of the same sort of testing that Genie had undergone. Here, in neurology, is where Genie's scientific importance had culminated; it was Neville's territory.

29

Within Neville's lab things are comfortable—floors are carpeted and windows look out over the La Jolla cliffs at a sungilded stretch of Pacific. But the contest of whimsy and industry continues: a model of the human brain, protruding plastic veins and all, shares a shelf with a copy of *The Berenstain Bears Go to School*. A scattered collection of toys vies with a blinking bank of computers, their monitors peopled with fairy tale decals. The electronics are a far cry from the bulky machinery Jay Shurley had to truck around to monitor Genie's sleep.

"There's a huge amount of work done on language at the Salk Institute, but the focus of my lab is the effect of experience on brain development and behavior," Neville told me.

"It used to be thought that genes determined how the brain develops. It's becoming clear that environment plays a certain role in setting up the organization of the brain, the specialization of its various parts. Is brain specialization intrinsic, or is it input? This is what we're trying to find out."

Neville spoke with me over a competing conversation between Eeyore and Winnie the Pooh. Pooh had been enlisted, in video form, to distract a ten-year-old subject whom Neville's assistant, Sharon Coffey, was readying for testing. The boy's name was Joseph. He was halfway through a cookie. His head was being fitted with a cloth helmet with ear flaps and chin strap—except for its lack of goggles, the sort of helmet Snoopy might sport in his incarnation as a World War I flying ace. This one, though, was made of brightly colored polyester mesh perforated with a number of grommets. Coffey squirted electroconductive jelly through each grommet with a large, blunt hypodermic needle. She appeared to be injecting Joseph's brain with petroleum jelly. With each application, he grimaced. "C'mon," Coffey said. "You've done this enough times. You know it doesn't hurt." Joseph kept his eyes on Pooh and continued to flinch theatrically. Babies are so afraid of the needle that the applications are administered by a hand puppet, Coffey told me. She attached seventeen electrodes through the grommets to Joseph's head, and when she had him wigged up like some electronic Medusa she led him away, leaving behind a chair full of cookie crumbs and Pooh with his head stuck in a honey pot.

In a small adjoining room, Joseph was seated in a ratty over-stuffed reclining chair facing yet another TV monitor. His electrodes were plugged into a larger wiring harness, secured with a strap around his chest. The room's walls were padded with burlap to deaden noise and shielded with lead so that the electronics would not pick up, along with Joseph's brain waves, the communications of submarines patrolling off the coast. When he was sufficiently wired and trussed, Coffey handed him for his amusement a sheet of dime-sized colored animal stickers. He had a pink brontosaurus plastered to his cheek before she could even get out of the room. She closed the armored and upholstered door behind her and, in the cramped antechamber, sat down at a computer console. From here she could control the images Joseph would see on his TV screen and the sentences he'd hear over his headphones. From here, also, she would receive reports from his bristling headful of electrodes, "eavesdropping," as Neville would have it, on the activity of the brain.

When the computers had warmed up, Coffey began the first test, a word recognition task where Joseph had to relate a word he heard with the image flashed on his TV screen. Sometimes the two agreed (key, key) and sometimes not (key, knife) and the recognition coursed through his head like an animated fish around a goldfish bowl. Outside the room, the computers traced the paths of swimming impulses. Seventeen jiggling pens scratched seventeen nervous, varicolored lines against a moving scroll of polygraph paper. The pens spasmed and Coffey spoke through a microphone to the boy inside the room.

"Joseph, you have to relax, OK? Relax your face. Don't hold your mouth tight like that. Just relax." And another time, "Joseph, don't get happy." She watched the polygraph paper and watched a monitor above it, where Joseph's little form showed up in grainy black and white, strapped in the chair in the little room, struggling with concentration. He gritted his teeth and Coffey said to Neville and me, "Stand back." The pens flew and ink splattered off the paper in an expectoration of red and blue and green.

Joseph, being a normal child, is considered by Neville to be among her most important subjects, for he provides a comparison for all the others—the children who are brain damaged, who are blind, who are retarded, those who are deaf and speak sign language and those who are deaf and do not, those who have had strange upbringings, and those with strange diseases. Neville had among her testees at that time one hundred normal children, and they hadn't been easy to find. "We can't run bilingual kids or children of alcoholics," she told me. "Their brain organization is different. Also, we always consider SES—socioeconomic status. There are a lot of kids who have had a stroke at birth. It's a common thing. If you are in a high SES, you may recover from the effects, but indications are that if you are poor, you will not."

Working with her "population" is as close as Neville has come to working with Genie, but it is pretty close at that. The "cycle test" Coffey was administering to Joseph is akin to the one devised by Susan Curtiss and used on Genie in the mid-1970s. In those years Neville was a doctoral student at Cornell. Her teacher was Eric Lenneberg. Her research experience began with animals, and she learned from them something of the strange power of critical periods. Cats' brains learn to perceive vertical lines between three and twelve weeks, and if during those weeks a kitten is presented with no models of verticalness, then the vertical lines it later encounters in the real world may appear invisible. If it is deprived of examples of horizontalness at that crucial time, it may ever after have the same problem with horizontal lines. "Also in cats," Neville told me, "the critical period for depth perception is from six to sixteen weeks. Beyond that, they will never get the brain organization necessary to see depth. Likewise, in some children, their eyes don't coordinate. When the eyes are not convergent, they must be corrected by the age of two or the child will never see depth. So the critical period for depth perception in humans is zero to two years old."

A wag of the nineteenth century observed that "attrition of minds, as well as of pebbles, produces polish and brilliance." His social critique would be accepted today as neurological verity. "Brain maturation is not about the way the brain grows," Catherine Snow had told me. "It's about the way it dies. As it ages, its neurons disappear." As the brain sheds neurons, it sheds its plasticity, its blank potential. But at the same time the shape of its character and skills is revealed, the way a sculpture is revealed by the chipping away of stone. The process is known as stabilization.

"Early on, the brain has more neurons and more connections," Neville said. "In the adult, there are no connections between the primary auditory cortex and the primary visual cortex; in newborns there are. There is also a transient connection between the retina of the eye and the auditory thalamus— I could give you lots of examples. These normally get eliminated. So what determines what gets stabilized and what gets eliminated? Those connections that get exercised are stabilized, and those that aren't, atrophy. Different functions stabilize at different times. Senses are stabilized early. Language is stabilized late. Critical periods are different for different functions."

By the onset of puberty all critical periods have run their course and stabilization is done. "The critical importance of puberty is something we don't understand," said Neville. "The hormones testosterone and estrogen are neurotransmitters and somehow they are critical in stabilizing the organization of the brain." Until that deadline the brain is flexible enough to compensate for problems. "For example," Neville told me, "brain damage can interfere with acquisition of language early on, but if it happens during the critical period, other parts of the brain can fill in."

Neville's subjects have taught her just how ingenious the improvisations of the unstabilized brain can be. "We learn a lot from deaf people who use American Sign Language," she says. "Facial perception is usually a right-hemisphere task. It's the right brain that knows from someone's face if they're sad or angry. Except that facial expressions are also important to the grammar of ASL, and grammar is left brain. We find that in deaf people grammatical facial expressions are perceived by the left hemisphere, but expressions of emotions are perceived in the right."

The deaf also debunked for Neville the possibility that the location of the language center might be a matter of convenience. Hearing humans are extremely adept at discerning different sounds, a skill that helps us to arrest those racing phonemes. The difference between "pah" and "bah" takes place in twenty milliseconds of speech, yet we catch it, and the others like it, almost unerringly. It had been proposed that language was piggybacking on this extraordinary auditory skill, that language, being primarily heard, resides in the left brain because that's where the perception of sound is governed.

The issue is one of territories. If King Psamtik's interest in the language question was geopolitical, the questions batting around Neville's lab are no less so. Only the terrain is different—Neville's hegemonies are more ancient than those of the ancient world. Her borders are those between contiguous states of mind. Which of our skills are vested, she wants to know, and which itinerant? Are the cerebral realms ruled by the senses and by cognition? Is language just a gypsy, resting wherever on the mental map it can make the best alliance?

Neville found her evidence in an unusual place: in the far corners of the visual field of people speaking ASL. The refinement of perception required of a speaking person's hearing is required of deaf people in a different way. They must be able to look a companion straight in the face and grasp the rapid grammar being delivered with a filigree of finger movements some distance to either side. As a result, ASL speakers develop exceptional peripheral vision. Would their language center have perhaps crossed over to cozy up to peripheral vision in the brain's right half?

Neville described to me the ensuing experiment: how she measured subjects' brain waves while positioning small squares of projected light around their field of view. Her enactment of the endeavor filled the electronics-cramped antechamber with gesticulation. Behind her, within the testing room, the cycle tests had ended. Coffey had put Joseph through sentence and general comprehension tasks of greater and greater complexity. As she unstrapped him from his chair and released him from the confining room, he was weary and restive.

"It turns out," Neville was telling me, "that, unlike the rest of us, ASL speakers see peripheral motion with their left hemisphere. Language tasks are *not* moved to the visual side. The visual ability is pulled into the language center. There is a strong biological constraint that determines that language is contained in the left hemisphere. Language is not dragged around!"

Joseph and Coffey pushed past Neville out of the lead-lined box.

"Did I do OK?" the boy asked, looking up. He had a blue seahorse stuck on the middle of his nose.

"Yes, you were perfect, Joseph," Coffey said. "You had perfect little brain waves."

30

From the earliest observations of Genie, it appeared that her brain function was biased. She was listing heavily to starboard. The tasks she performed well were all right-brain tasks; the tasks she failed were left-brain. Genie's response to tasks requiring an equal collaboration between hemispheres was frustrated and hesitant, with none of the quick confidence she displayed when thinking "right."

The dominance of one hemisphere or one lobe in any given task is never total. Both sides of the brain work on every task, but their collaborations are lopsided. How the tasks are divvied up depends on the individual—the male's brain organization may differ from the female's; some researchers believe that women are less lateralized or perhaps have more connections between their hemispheres. Also, some left-handed people have the entire mental map reversed, so that language is on the right and visual ability on the left. In the fine points of brain layout, we are each of us different from our neighbors.

Genie's deviation, however, was extreme, and Curtiss wanted to know why. Her opportunity was provided by another aspect of brain physiology. Each side of the brain controls the opposite side of the body, so that the right brain, for instance, moves the left hand. Unfortunately for neurolinguists, you cannot whisper to the left brain through the right ear without the right brain's overhearing you, because each ear is wired to both sides of the brain. The connection to the opposite side is stronger, however, and in one circumstance it has a near monopoly: when a sound is presented to the left ear at the same time that a different and competing sound is presented to the right ear, each ear reports almost exclusively to the opposite side of the brain. This oddity makes possible what is called the dichotic listening test. By playing different sounds simultaneously into each of Genie's ears, Curtiss was able to speak directly to each hemisphere of her brain and measure each hemisphere's response.

"What matters is the material the ear hears," Curtiss told me. "Language is handled better by the right ear, and environmental or musical sounds by the left ear. We played environmental sounds to Genie and checked her response. Each ear alone performed perfectly; both ears with the same sounds were OK; but when the two ears competed, the left ear performed better. That's normal—but the degree of the asymmetry was not. Then we fed her words the same way." The results bore out long-standing suspicions. Genie's brain was processing language just as it did environmental sounds—on the right. The right brain was handling work usually done across the aisle. The real surprise lay in the degree of the imbalance. Normally, the dominance of one side over the other shows up in the dichotic listening test only as a subtle preference—nothing too pronounced. With Genie, it wasn't just pronounced, it was absolute.

Seeking a second opinion, Curtiss took Genie to the Brain

Research Institute, on the UCLA campus. "We attached elec-trodes to her skull to read her brain waves as we showed her pictures or read her sentences," Curtiss told me. "After each stimulus like that, you will get a response in the form of a wave. You will get a seeing event as she spots the image and then, three hundred milliseconds later, a recognition event. First, we showed her faces. Her response pattern was parallel to the environmental-sounds test—that is, the right hemisphere showed a greater response than the left. Normal. Then we played sentences." The results, as before, were extreme. Genie's performance was as lopsided as that of children whose left hemispheres have been surgically removed. She didn't seem to be using her left brain for language at all. When it came to its central function, her left brain was functionally dead.

"Why should this be so?" Curtiss asked in a paper on lan-guage and cognition published in *Working Papers in Cognitive Linguistics* in 1981 and later reprinted in other journals. She continued, "Genie's case suggests the possibility that normal cerebral organization may depend on language development occurring at the appropriate time." To the question "How do we acquire what's innate?" Genie was suggesting an answer. As Catherine Snow once phrased it: "Lenneberg claimed that the brain organized language learning. Now it seems certain that stimulation is needed to organize the brain." Snow and some other linguists hold that any logic system might suffice as a stimulus: music, for instance, or the logic of societal interac-tion. But Curtiss read her data differently: only language would do.

Curtiss had run her finger down the string of Genie's expe-rience until she encountered the fabled, elusive knot—the tie between language and humanity—and found that knot to be more concrete than Itard or Sicard or Condillac could ever have suspected. If Genie was any indication, we are physically formed by the influence of language. An essential part of our personal physical development is conferred on us by others, and comes in at the ear. The organization of our brain is as genetically ordained and as automatic as breathing, but, like breathing, it is initiated by the slap of a midwife, and the midwife is grammar.

A slap is all that's needed. "It seems to take a phenomenally small amount of input to trigger this special process," Helen Neville told me. To support her observations of Genie, Curtiss cited Neville's experiments at Salk, but that in itself posed a problem. Neville's relevant work was on a population for whom nothing came in at the ear—deaf people speaking ASL, a "tongue" so different from English on its surface that it is hard to see how they could both be considered language. Fortunately, however, that problem had already been dealt with, by another researcher working one floor down from Neville at the Salk Institute.

31

Ursula Bellugi had been at Salk since the late 1960s, and in 1971 she was called to Los Angeles to meet Genie and confer about her fate. She is a person who confronts nothing except head-on, a carrot-haired fireball of renegade enthusiasms. On the morning I first met her she was wearing a royal-blue dress with a blue leopard-spotted wrap and, for jewelry, a large seashell suspended on a cord. The shell was as tossed about by the heaving blue storms of Bellugi's conversation as it had ever been by waves. When her rising tirades brought her to the threshold of comprehensibility, Bellugi stood at her desk and implored herself, "Hold on, Ursi! Hold on," and forced herself back into the more languid speaking pace of, say, a tobacco auctioneer.

"Jonas Salk asked me if I would come set up a lab that would apply biological methods to the study of language," she said. "I soon settled on ASL as a way to study language acquisition in general. Then the question arose: Is ASL a language? Or just a universal pantomime or a broken English on the hands? Is there any structure at all to what deaf people do when they communicate? We had to grapple with the question: What do we mean when we say that something is a language? This was a great voyage of discovery. What does it mean to say something is a language? I asked the question every night, and every night I loved it."

Whatever else it might be, ASL is definitely not an English of the hands. If it has any spoken kin, that would have to be French, for ASL's roots are in Paris at the Institut National des Sourds-Muets during the era of the Wild Boy of Aveyron. It is a branch of the sign language devised by Abbé de l'Epée and taught by Sicard to the deaf freed from the Bicêtre. In 1810 it was brought over to Connecticut, straight from the institute, by the American educator Thomas Hopkins Gallaudet. Despite that estimable history, ASL had never been seriously scrutinized, at least not with Bellugian rigor. At Salk, Bellugi investigated ASL wit and

ASL metaphor. She found a deaf poetess and studied ASL verse. She found ASL intonation, couched in the subtle modulation of hand movements. She catalogued slips of the hand and the fingered inflections of a deaf Asian whose ASL was accented with Chinese Sign Language. And with Ted Supalla, a deaf linguist now at the University of Rochester, she vivisected ASL's hidden structure and found within it essential characteristics of syntax, semantics, and phonology.

"It's the same as spoken language," Bellugi concluded. And importantly, it is organic. It changes, as do all true languages, through a constantly shifting series of local forms. Linguists like to amuse themselves with their sensitivity to those forms, as I discovered when I interviewed Elissa Newport, who studied with Lila Gleitman and now teaches at Rochester, and who is married to Ted Supalla. She and I had talked no more than ten minutes when she turned to graduate student Jenny Singleton and asked, "Well?"

"Georgia," Singleton replied. "But he didn't stay long before moving to Virginia, somewhere on the coast, and was in New Jersey before he was four, but returned to Georgia." And she proceeded to sketch out accurately the rest of a rather fugitive childhood.

Had I been an ASL speaker, my early itinerary would have been all the more transparent, for ASL dialects are unusually distinct. The language lives in small populations centered around schools for the deaf, such as Gallaudet in Washington, D.C., and the California School for the Deaf in Fremont, California. The populations are isolated. There are no pervasive media like radio or television to impose on ASL a damping national conformity. At the Michigan School for the Deaf, the popular off-campus hangout for a generation of deaf students was a store with a cigar-store Indian out front. "Meet you at the Indian" became the local vernacular—and now, if you are speaking ASL with anyone anywhere in the world and they use the sign for "Indian" as a synonym for "store," you can be sure that they hail from Michigan.

The same sort of evolution occurs with all formalized languages, and the innovations are generally regarded by the educated as corruptions of the mother tongue. It would seem strange to hold up such corruptions as evidence that a language is bona fide. Should the Académie Française truly be ecstatic about argot? Does ghetto slang attest to the viability of English? Evidently, says Catherine Snow. "Remember what Mark Twain said," she once admonished me. "'A language is a dialect with an army.' There's a bit of a political problem when analyzing ghetto speech. How do you know if someone doesn't speak well, or if they just don't have an army?"

For Bellugi, the changeability of ASL displayed its true vitality—it lived in the street, not in the academy; it was owned by its speakers, not dictated by its experts; it had surpassed its invention and was home grown every day. Bellugi and Supalla became ASL's army. It was a true language, they contended, as organic and robust and intricate as any.

The certification of ASL was an enormous gift to the armamentarium of modern linguistics. There has arguably been none greater. Sicard, in the time of Victor, and Abbé de l'Epée before him, were among those who bequeathed to our generation the question of the relationship between man and language. In devising the roots of ASL, they had conspired with Ursula Bellugi to provide, for the riddle, a key.

The key's effectiveness is due partly to the history of the deaf. Despite their symbolic liberation by Abbé de l'Epée, the deaf still exist in a world of chronic misunderstanding. They are often misdiagnosed as retarded and languish in misdirected programs. Even families that understand the condition of deaf children may feel that they would be better off learning to read the speaking world's lips rather than the hand signs of a foreign-seeming and insular culture. Thus the deaf may have contact with their first real language at two, or at five, or at fifteen years of age. Their plight has provided linguistics, in recent years, with a hundred Genies, and what's better, with Genies who have not been psychologically abused, only linguistically deprived.

The deprivation can be marvelously inventive, and the inventiveness enlightening. At the University of Illinois, Jenny Singleton studied something called the pidgin and Creole phenomenon. Her subjects were a deaf family isolated in a hearing town. The parents spoke ASL, but they had learned the language late and their grammar was poor. Their deaf son was exposed to their pidgin ASL from birth; it was his only linguistic input. Yet when Singleton analyzed the parents' speech and the son's, she found that the boy had developed proper ASL from the flawed model. He could speak better than anyone he'd ever heard. His virtuosity supports Chomsky's contention that syntax is biologically, not parentally, provided.

A lot of deaf people filed through Neville's lab at Salk in 1977 and 1978. She performed lateralization tests on a large population, and this is what she found: The deaf who learned ASL during childhood had left brains lateralized for language as well as for other tasks. But those who were deprived of early sign language did not—their brains were yet unformed. The midwife had not spanked the baby.

"Relating Neville's data to Genie's case suggests that language development may be the crucial factor in hemispheric specialization," Curtiss wrote in her 1981 paper. "When [language] develops, it determines what else the language hemisphere will be specialized for. In its absence, it prevents the language hemisphere from specializing for any higher cortical functions." The insight promised to redefine some basic intertwined ideas: What does it mean to say that something is a language? Language is a logic system so organically tuned to the mechanism of the human brain that it actually triggers the brain's growth. What are human beings? Beings whose brain development is uniquely responsive to and dependent on the receipt at the proper time of even a small sample of language.

In the light of all this, then, what was Genie?

32

Curtiss's best attempt to grapple with this question remains her doctoral thesis. It is the most significant published result of all the research on Genie—significant enough to be cited in virtually every current American textbook on basic linguistics, sociology, or psychology. In addition, it was picked up for publication as a book, something rare for a scientific thesis. *Genie:*

A Psycholinguistic Study of a Modern-Day 'Wild Child' was published by Academic Press in mid-1977. Besides sporting hard covers, the book differed from the dissertation in having a dedication page, which read "To Genie," and a frontispiece, which was a pencil drawing of a smiling person with curly hair and big ears holding a small figure in its left arm. Curtiss's caption for this drawing read, in part:

> Early in 1977, filled with loneliness and longing, Genie drew this picture. At first she drew only the picture of her mother and then labeled it "I miss Mama." She then suddenly began to draw more. The moment she finished she took my hand, placed it next to what she had just drawn, motioning me to write, and said "Baby Genie." Then she pointed under her drawing and said, "Mama hand." I dictated all the letters. Satisfied, she sat back and stared at the picture. There she was, a baby in her mother's arms. She had created her own reality.

In late 1977 and early 1978, as *Genie* the dissertation succeeded to *Genie* the book, Genie the person continued her tribulations. The home she had entered after her two-week stay at Childrens Hospital worked well for a while but then fell suddenly apart, victim to the foster mother's family pressures. Genie was lodged elsewhere temporarily during the Christmas holiday, then placed in yet another, supposedly permanent, home. Curtiss continued to agitate for her friend's welfare. "Genie is confused and traumatized by these frequent moves," she wrote in a letter to John Miner on January 6, 1978. "Not only because these moves are unsettling and require repeated readjustments, but also because she feels continually rejected—that somehow she is forced to move each time because she has been a 'bad girl.'" Curtiss was especially distressed because the newest of Genie's homes was adamantly refusing visitation to the people who had figured so large in Genie's life during her happy years at the Riglers. Curtiss had always been foremost among those visitors. Although she did not know it when she wrote the letter, Curtiss had already seen Genie for the last time, three days earlier.

The moratorium on visits was one of a host of ongoing petty skirmishes and diplomatic failures between the old Genie Team and Genie's new caretakers. Susan Curtiss had not been the only one fretting about Genie's "abandonment" after she'd left the Riglers. The social workers overseeing Genie's case for the DPSS and the regional center complained that they were having trouble reaching John Miner when they needed his signature as guardian. Their letters to his office often went unanswered and their phone calls unreturned, with the result, they said, that critical authorizations were being delayed and Genie's welfare jeopardized.

A year after leaving the Riglers, Genie had found herself in particularly dire financial straits. The Social Security Administration had discovered her one-third interest in the house on Golden West Avenue; the agency rescinded her eligibility, cut off the payments that had been supporting Genie in her foster home, and demanded she repay them $1,230 for benefits she had already received. Genie's social worker appealed to Miner for help but was told that the $4,000 being guarded in Genie's trust fund was not hers to use either for her care or to satisfy debts. It was owed in its entirety to David Rigler, Miner said, for services he'd performed for Genie while she was in his home. The social worker, Thomas Greenan, was outraged, and after a visit with Rigler in which Rigler said he felt he was owed the money, Greenan made his umbrage official. In October 1976, Greenan filed a complaint with the Probate Investigation Unit of the Office of Public Guardian, in which he accused Miner of inadequately attending to Genie's welfare. He requested that guardianship be transferred to the East Los Angeles regional center.

Miner, for his part, considered Greenan and the other social workers bureaucratic obstructionists playing a game of jurisdictional one-upsmanship. He refers to employees at the regional center as "miserable sons of bitches of the first water" and its functionaries as "horses' asses." He felt the agencies were spiteful and uncooperative, complicating rather than helping what efforts he did take on Genie's behalf. Miner contested the social security ruling and was eventually able to get it reversed, by pointing out that Genie's house was not a liquid asset and so should not be counted among the tangible riches that would disqualify her for benefits.

For their part, the workers at the DPSS and the regional center felt strongly that Miner's inaction on the money question had abetted the breakdown in Genie's first foster home—thus contributing indirectly to Genie's regression and rehospitalization—and had clouded further placements as well. They persisted in their efforts to replace Miner as guardian. A lawyer they hired to investigate the matter, Janice Stone, reported that Miner's inattentions had left his own role vulnerable—he had failed to change Genie's guardianship over from that of a child to that of an incompetent adult when Genie turned eighteen in 1975. Therefore, the guardianship had officially lapsed. Without consulting Miner, the regional center had the guardianship transferred—to Genie's natural mother. On March 20, 1978, Irene at last regained control of her daughter.

33

Irene was not yet, however, in control of Genie's estate, simply because Miner would not relinquish it. In the face of Janice Stone's repeated petitions and the court's repeated requests that Miner appear and show cause why he would not pay up, Miner responded with an old claim. He presented a bill against the estate from David Rigler, for psychotherapy Rigler had provided Genie almost four years earlier, during the first six months of 1975. The rate, $50 per hour, had been agreed upon in a conversation between Rigler and Miner in the summer of 1971. Miner claimed he had been acting then as Genie's guardian; in court, however, he was reminded that he had not been appointed guardian until the following year.

In the court hearing, held in November and December 1978, David Rigler gave a description of the therapy he'd supplied. "[It] was rendered under naturalistic conditions," Rigler testified. "It was not rendered in the conventional office psychotherapy context."

Stone asked Rigler to explain "naturalistic conditions."

He answered, "Surely. Genie was living in our home. She would be taken to school. Sometimes she took the bus, sometimes she would be taken. She would be taken on shopping ventures. She would be taken deliberately out for walks. She would be taken into the backyard and worked with there in the context of a natural living situation. A great deal of this kind of treatment was provided in this natural context."

Rigler had not kept any documentation of these therapy sessions and had presented no itemized invoice. But he had communicated to Miner that, by his estimation, his efforts had accrued to the tune of $7,800. The court inquired about the Riglers' other reimbursement while Genie was in their home. It heard of Marilyn Rigler's stipend, under the grant, of $500 to $1,000 a month, and of David Rigler's release with pay from Childrens Hospital, and of the foster home support the Riglers had received. That support had climbed, by 1975, to $552 a month, not prorated for Genie's spending a third of her time at Irene's. In addition, both Riglers had profited from articles and speeches.

"The Riglers have indeed been compensated many times over," Irene's lawyers contended. They also brought up a practical problem: there was not enough money left in Genie's estate to satisfy Rigler's bill. This had already occurred to the petitioners, who were asking for only $4,500. As Miner testified, "[Rigler] simply said, 'Well, whatever the estate can afford will be agreeable to me.'"

"Did he suggest the figure of $4,500?" asked R. Samuel Paz, another of Irene's lawyers.

"Well," Miner replied, "that was all there was in the estate. So whatever the estate could pay, he was willing to accept."

In the end the court was swayed by testimony from Rigler and Miner concerning the "substantial benefit to the child" of her being in the Riglers' home: how her vocabulary had

advanced, how she had been toilet trained and taught to interact with the world. The gains had been hard won, Miner told
the court. "From my own observation it was clear that Genie
presented unprecedented problems, because she is an unprecedented case, to the Rigler family. It involved them twenty-four
hours a day for four years. She had adjustments to make, and
the process of attempting to train her and to work with her
resulted in her becoming very distressed, throwing tantrums,
being extremely difficult to manage, and . . . it was a matter of
my not being able to understand how people not related to this
child could undergo what, in fact, she was subjecting them to
in terms of the strain on the household."

"If I may add one thing with respect to this improvement,"
Miner said at another point. "Genie's demands on the Rigler
household were such that I could not believe that any people
would subject themselves, except out of great love and compassion, for what they did for Genie."

It was a testament to Rigler's effectiveness that by 1975
Genie had advanced to where she was able to return home to
her mother, the petitioners said. It was also a testament that
since leaving Laughlin Park, Genie had undergone "a great
deal of deterioration."

In giving his decision, the judge admired Rigler's ministra-
tions. "These were services which brought this child back
from, well, a nonexistence, a life as an animal and to the place,
in part, where the child could function in society. Apparently
not much. The child is apparently quite limited mentally and
can't develop perhaps beyond a certain point. But, in any case,
he did bring the child a considerable distance away from her
original status, which was a nonstatus, I would say.

"Yet," the judge continued, "we're dealing with an estate which is quite modest." In light of that, he knocked the award down to $3,100, including $600 earmarked for the petitioners' legal fees.

"I didn't do this for the money. I never had funds in mind when I took Genie in," Rigler told me when I first asked him about his claim against Genie's estate. He was slumped into his leather office couch looking uncomfortable and mildly forlorn. His memory of the claim was fragmented and adamant. It was Miner's idea and not his, he said. He had never seen any money from it. He didn't know if Miner had received the money. And anyway, they had intended to put the money they received into a trust fund for Genie. In subsequent conversations, he elaborated, saying that Miner had been afraid that if he released the money, it would not go to Genie at all but to the state. Miner hoped that by sequestering the money, it could be kept for her use when he was able to retrieve her guardianship. Having no unassailable claim on the funds himself, Miner enlisted Rigler into petitioning for reimbursement. Rigler had collaborated, he told me, on the condition that Miner establish a fund for Genie with the money they gained. "I didn't really believe the things I was saying," Rigler told me of his court appearance. "I heard myself saying them, but I didn't really believe them." He gave a snort that seemed part rueful laugh, part relieved sigh. "I've never expressed that before to anyone," he said.

John Miner confirms Rigler's version. In a letter he wrote me responding to my questions about the case, he cautioned that his files on Genie had been lost in an office move, and that his memory was vague. He stated of Rigler's claim: "It was my idea, not his," and continued, "The notion that the service Dave Rigler rendered to Genie was financially motivated is despicable. Certainly it was a unique professional opportunity. But let no one forget Dr. Rigler damn near sacrificed his family on the altar of helping Genie."

Both men insist that the claim, though granted by the court, was never satisfied, and that no money was disbursed from Genie's trust fund to Rigler or to Miner. Miner explained to me that he could not, himself, have satisfied the claim, because he "had no minor's estate fund to pay Dr. Rigler." I later showed him a document (which he said mystified him) indicating that he still, in fact, had control of Genie's estate long after the court decided Rigler's request. The document was a petition, signed by him and filed with the clerk of Superior Court in July 1979, requesting removal of Irene as guardian, claiming that she was unfit. In it, Miner volunteers to be reinstated as guardian, offering to serve, as he had before, without fee, and notes that he is still holding in trust for Genie $2,126.02, the amount remaining in her estate after the court order, of half a year earlier, settling Rigler's claim.

The removal petition was denied, and Irene remained guardian, but it would take another year and a half and several more court sessions before she could wrest Genie's estate from John Miner. When she finally received the check, in February, 1981, it was minus the money supposedly disbursed to Rigler. Rigler and Miner both say that they do not know what became of that money.

34

In the long run, the true costs of the wrangling over Genie's guardianship went far beyond dollars and cents. I once asked Rigler if he thought Irene's animosity toward the scientists might have been exacerbated by his suing for Genie's inheritance. "It's possible," Rigler allowed, and nodded. "I hadn't thought of that."

But the biggest provocation for Irene was the copy she had received, in early 1978, of Susan Curtiss's book. Her reaction was apparently instantaneous—she disliked it even before she'd opened it. "When I saw the title of the book, I felt hurt," Irene wrote. "My daughter . . . classified as a 'wild child.'"

Irene's rebuttal was handwritten on lined loose-leaf paper and was addressed "To Sam"—R. Samuel Paz, who, along with an attorney named Louise Monaco, would act as Irene's lawyer in the long legal season that was about to ensue. Irene's letter protesting the dissertation became Exhibit B in that season. Exhibit A was the dissertation itself.

Irene was especially incensed at Curtiss's opening chapter, which recounted Irene's life with Clark and the dreadful tribulations of their children. In a sworn deposition she gave several years later, Irene expressed her reaction to that material. "It made me sick at the stomach," she said. "I was sick, you know, when I saw those things, you know, in print." She added, "It takes a lot to make me sick."

In her letter to Paz (in which she calls her daughter by her real name, which I have replaced), Irene quibbled with much of Curtiss's description. Irene wrote:

- I was not frequently beaten. 2 times in the last year. He did try 1 time to kill me. . . .
- Genie was never forgotten and I did the best I could in taking care of her. . . .
- It depended on the weather to what she wore while sitting on the potty chair. She was able to move her arms, legs, bend forward and to the sides.
- [Curtiss] writes as though Genie stayed all the time on the potty chair.
- Genie was never forgotten.
- Genie was able to move her arms when she had her sleeping bag on. It was not a straitjacket. It was an oversize infant's crib with wire screen around sides. There was a wire screen top but I never used it. . . .
- Genie did hear speech.
- Our home is very small. . . .
- She could hear the traffic noise from street.
- She heard the neighbors next door coming and going. . . .
- She heard airplanes, birds, neighbors, traffic noises.
- Genie was not forgotten.
- Her father did not beat her.
- The paddle was not left in Genie's room.

- Her father did talk to her.
- Once in a while he did bark at her to distract her making noise without opening door.
- He never barked at her face to face.
- He talked to her.
- He did not scratch her. . . . He did not beat Genie.
- He did not stand outside of her room and bark and growl at her. . . .
- There was a chest of drawers, a chair, a folding bed, 2 large trunks, window shades, and curtains. Oversize baby bed. Potty chair.

Irene's official complaint was not about inaccuracies. It was, rather, the opposite—that depictions as detailed as those related by Curtiss, and by other scientists in various papers and speeches, could only have been pilfered from Irene's own privileged conversations with her therapist, Vrinda Knapp, and with Knapp's supervisor, Howard Hansen. In October 1979, Irene filed suit in Superior Court against Hansen, Knapp, David Rigler, James Kent, Susan Curtiss, and Childrens Hospital, accusing them of multiple infractions of patient–therapist and patient–physician confidentiality. The defendants had, the suit claimed, "exposed, revealed, and published to the public personal, confidential, and intimate details of the years of imprisonment, suffering, isolation, abuse, and torture" suffered by Irene and Genie.

That wasn't all, or even the worst. The fourth of five causes of action in the suit accused the scientists of subjecting Genie to "extreme, unreasonable, and outrageous intensive testing, experimentation, and observation" under "conditions of duress and servitude"—in short, of performing unethical human experimentation. The remaining cause of action faulted John Miner, for not protecting Genie from harm while he was her guardian. Irene asked for both compensatory and punitive damages.

Sam Paz was well prepared for the issues in the case, scientific as well as legal. When he was an undergraduate at UCLA he had majored in psychology and had trod some of the same intellectual hallways as did Victoria Fromkin and Susan Curtiss. "At one point, I went through Curtiss's book and tallied up the experimenting that was done," he told me. "The intensity and frequency of sessions was high. There were other research papers, too, and if you look through them you will get a good idea of what Genie had to endure. She was on a testing regimen, at one point, of sixty or seventy hours a week. The response when we asked the researchers about this was that it was fun—that Genie thought of most of this as a game."

Susan Curtiss decries the idea that her testing was ever so demanding. "My test periods were never more than forty-five minutes on a given day," she told me. "The rest was playing, going for walks, just being friends." It was a point she emphasized in sworn depositions, along with her impression that Genie had enjoyed the work. "She loved the social praise and social reward, which I always gave her for just responding, whether she was right or not," Curtiss testified. "The time that we spent testing was a time when usually it felt intimate. It was just Genie and me very close together, a lot of eye contact, a lot of—we would hug and cheer and do a lot of things, and no one else would—I mean, we wouldn't have crowds of strangers there that had to be contended with at the same time. So there were things about it that I think were very meaningful to her. And she always got tangible rewards at the end, or even— either a tangible reward or a reward in terms of some special treat that we would go and buy and go and do together, and she loved that part of it, too."

Curtiss also noted that testing was sometimes done at Genie's insistence. Her testing of Genie had ceased entirely during 1977, as her work under the National Science Foundation grant turned its focus to other children whose conditions could cast further light on the lessons learned from Genie. Curtiss was shocked by the news of Irene's legal action, as were the other scientists.

"The suit was right out of the blue," Rigler told me. "One Sunday morning, we got a call from a friend who said, 'Did you know your name is in the paper?' So we got the *L.A. Times,* and that's how we learned we were being sued. And it had Genie's real name, and we'd been so careful all those years to keep that away from the public." Rigler and Hansen both had the immediate impression, faced with the charge of human experimentation, that they were going to be taken to task for the sleep experiments conducted early on by Jay Shurley, since those studies had in some ways been the most physically invasive. They were wrong—Shurley, in fact, had been contacted by Irene's lawyers and had offered to testify for the prosecution and against his former colleagues.

However ample its inspirations, the suit seemed remarkably adventurous, coming from a woman who was described even by her lawyers as a timid individual. David Rigler remembers the moment when the mystery was made clear to him, the hidden hand revealed. "When I gave my deposition, Irene's lawyer had a copy of Curtiss's dissertation marked up, with passages underlined that were supposedly slanderous of Irene," he told me. "I asked if I could see the book, and he handed it to me, and the front cover fell open, and the name written inside was *Jean Butler Ruch*."

In the eight years that Jean Ruch had been Rigler's unflagging antagonist within the scientific community, Rigler had had no suspicion of her growing association with Irene. By Ruch's account, that association had suffered a hiatus of four years, after Irene called her one afternoon to cancel a visit, saying that Rigler had forbidden her to see Ruch under penalty of losing visitation rights with her daughter. When the Riglers were no longer Genie's gatekeepers, the mother and the schoolteacher were emboldened to find in their common antipathies the grounds for an alliance.

"Ruch stayed in the shadows, but she was constantly chiding Irene—putting a bug in her ear that the scientists were overreaching," Paz told me. "Her involvement seemed to be the catalyst. My own assessment is that Irene was very passive, that she would never have done this on her own. When she called me, I felt that I wasn't really talking with her but with Mrs. Ruch. She wouldn't sound like herself, she would be very assertive. 'I want to do this!' or 'I know what's going on!' I didn't get the feeling that I was dealing only with Irene."

"Ruch was a sort of Svengali for Irene," Louise Monaco told me. "She was a major manipulator, as far as getting the lawsuit going. In the midst of all the power struggles, she was furious that Rigler had aced her out. She started the whole notion that Curtiss's book was a violation of Irene's privacy."

35

In length as well as rancor, the court case proved epic: the process of first and amended complaints, motions to compel, discovery, demurrers, depositions, and judgment stretched out over six years. In its course, the case provided plenty of fuel for the outrage of Irene and Ruch. In one early deposition, Howard Hansen stated that the records of Irene's psychotherapy, which contained information so sensitive that they were not allowed out of the psychiatric ward, were lost entirely, gone without a trace.

But in general, time dulled the sharpness of the complaint. The longer the case dragged on, the stronger grew the suspicion on the part of Irene's lawyers that they were contesting marshy ground. The same endless recitation, in Curtiss's book, of test procedures and test results, which had given rise to the charges of human experimentation, made a mockery of the notion that Curtiss had intended her dissertation as a potboiler, that she had exploited Genie's sad past for the sake of profit. From the start, Curtiss had wanted Genie to receive any profits her dissertation earned, and had placed them in a trust fund she had set up in Genie's name. In the alternating climates of complacency and controversy, Curtiss had found no one interested in helping her in her plan, until Paz and Monaco seized on her generosity as a way to break the deadlock in court. They recommended to Irene that she accept the money.

"We got to the point of settling the case in what I thought were the just interests of Genie," Paz said. "But Irene was prodded by Jean Ruch to decline Curtiss's offer. Ruch thought that it was unsatisfactory—that Irene should receive a lot of money. But the privacy issues related to Irene just weren't that strong. She had become a public figure."

Faced with Irene's intransigence, Paz and Monaco withdrew from the case. It was to be decided in chambers, and Irene went before the judge representing herself. It was now 1984, and the principal characters were subtly (or not so subtly) changed from those who had been there at the start. Floyd Ruch had died of bone cancer in 1982, leaving Jean a widow. Susan Curtiss, now Dr. Curtiss, had married and had given birth to her first child. Paz had become president of the Los Angeles ACLU. Owing to "economic exigencies," Childrens Hospital had undergone a severe bloodletting, and the department of psychiatry had been virtually eliminated. James Kent had moved to Children's Institute International, a child abuse treatment center, and David Rigler had opened a small private practice in northern California. He had to drive back to Los Angeles to sign the settlement papers.

Irene's complaint was essentially dismissed—or, rather, upheld, in a Tom Sawyerish bit of jurisprudence. The things that Curtiss had wanted to do with Genie she was now instructed to do by the court. She agreed to direct a program for Genie of linguistic, neurolinguistic, and neuropsychological evaluation and language instruction. Childrens Hospital was enjoined to give Genie yearly physical and psychiatric evaluations. To fulfill such obligations, Curtiss and the other defendants had full access to and use of Genie's records, and were granted the use of Genie's family history in scientific publications and speeches as long as they observed certain modest proprieties and donated any income to Genie's estate. As a first step in that direction, Curtiss relinquished the fund she had set aside for Genie, containing $8,383.79, her royalties to date from her dissertation. No other financial penalties were imposed.

In September 1984, David Rigler addressed a letter to Irene. "After what seems like forever, there is no lawyer telling me that I may not write to you. I have been wanting to write or talk to you for a long time," he wrote. "There are many things that I have wanted to say. . . . I do not know how you will feel about reading a letter from me, even whether you will read it. It is more than five years since we talked, and as a result of all the things that have happened in those five years, you may believe some things about me that I may feel are not true."

Rigler expressed his good will toward his courtroom adversary ("I have never felt anger toward you") and asked her to consider his rendition of the events of the last decade and a half. Coming after so many years of detailed depositions and sworn testimonies, the rendition seems almost mantric in its restatement, as though, by one more repetition of the verse, the resolving chord could be found.

"When Genie was admitted into Childrens Hospital in December of 1970, none of us could have predicted all of what would follow," Rigler wrote. "All any of us could do was to try our best. In July of 1971 she was released 'on leave' to

Mrs. Butler, and all of us who saw how she did out of the hospital wanted to keep her out. While Mrs. Butler and I disagreed about what might be best for Genie at that time (and we disagreed strongly), everyone (including Jean Butler) wanted the best for Genie. I don't think that Mrs. Butler will ever believe that I did not have anything to do with the fact that her home was not licensed to keep Genie, and that we ended up taking her only because there was not another good home available. But it was never our intention to keep her for ourselves, and during the nearly four years she lived with us, you will remember that we worked with you, knowing that Genie loved you. So it was always our idea to help you and Genie get back together. Of course, we were disappointed when it was necessary for you to place her in the [foster] home, but Genie was never easy to care for. Not for us, not for you, and not for anyone else. . . .

"We started work with Genie as a patient in the hospital. Before it was over, we got to live with Genie, and came to love her. If there is anything I regret, it is that we did our best and it was not good enough to restore Genie to be the whole human being we wanted (and still want) her to be. It is possible, I suppose, for someone else to have done it differently, but I think we did as good a job of it as anyone could. I am sure that Jean Butler and Floyd must have felt that we were doing everything wrong and they could have done it much better, but I have not seen anyone else do any better. Jean also seemed to feel, I thought, that Genie was not happy while she lived with us. If so, she was wrong. Genie was happy with us, and she came to love us and our family. . . . (I do think she missed contact with Jean, who I believe also loved her. I think it would have been nice for her if she could have kept up some contact with Jean during the time she lived at our house, but we were told that we should not permit that, and we had no choice.)"

The heart of the letter was a plaintive suggestion. "One of the difficult things during these years is that we have not had any communication with either yourself or Genie, and we have often wondered how you are getting along, and where Genie is, and what has been happening with both of you. We have also wondered whether Genie has ever asked or talked about us. I know we would like to see her if that is possible, some time when we visit in Los Angeles."

The Riglers' hope fell on unsympathetic ears. "Dear Dr. Rigler," the return letter read. "I'm very busy so this will be a very short letter. I am Genie's guardian. I do not wish to see you. You do not have my permission to write about Genie or me or my family." It was signed by Irene.

The humiliation Irene felt she had suffered at the hands of the court did nothing to assuage her anger. She ignored the settlement's condition that she not deprive the scientists of access to her daughter, and she hid Genie away. Genie currently lives in a home for retarded adults, and visits her mother one weekend each month. With the exception of Jay Shurley, none of the scientists has seen her. They do not know where she is, nor, except for rumors, have they heard how she is doing. In 1987, Irene sold the house on Golden West Avenue. She left—for the scientists, at least—no forwarding address.

V

THE WORLD WILL NEVER UNDERSTAND

36

Not long ago, I paid a visit to David and Marilyn Rigler in their new home, a pretty, two-story frame house on the northern California coast. The house was smaller than their previous one, but it didn't need to accommodate the life they had led in Laughlin Park: the children were grown, the Steinway was sold, and Tori's ashes were spread beyond a windbreak of eucalyptus in a field across the road. Genie remained only in a voluminous collection of reports, films, drawings, and photographs squirreled away in the back of the Riglers' garage.

We sat in his office, a downstairs room so strewn about with papers, books, old tape recorders, and film projectors that it seemed more the reliquary of a career than a place where one might still be carried on. There was a leather couch and a gray metal desk, and on the wall, amid the diplomas and citations, a print that seemed an odd choice to grace the office of a therapist. It was the optical illusion by M. C. Escher of an endless circular stairway that goes nowhere and only seems to ascend.

Rigler was in his late sixties, burly, gray-haired, and marked by an air of gentle domesticity and an expression of earnest and distracted kindliness. He described his feeling about the telling of Genie's story as "discomfort" and, later, as "dread." But to the degree that he was not reticent he was often confessional. Though he was too jealous of his documents cache to let me peruse it, he made repeated trips to the mysterious garage to drag out paper after video after drawing.

Some of the worst criticisms I had heard of David Rigler were from people he almost insisted I go see. "Has anyone mentioned Jay Shurley to you?" he asked. "When the fallout came, he had formed a very nice relationship with Irene and went over to her side. I don't know if he gave a deposition against us or not." And another time, "You must talk with Sam Paz. I always thought he was a man I would have liked, if he hadn't been suing me."

"Understand," he told me when we turned our attention to the years he had been the principal investigator in the research done on his foster daughter, "no one ever came to me and said, 'Dave, you should be doing X, Y, and Z'—except for Jay Shurley, who came in with a philosophical point of view. From his work with isolation cases, he said, 'You've got to let up on the pressure gradually, as though you had someone with the bends and you were bringing them to the surface. Let her come out a little at a time.' That had an impact on me. It was a useful notion. I don't think Shurley ever understood how much I tried to use his ideas."

Rigler stared at his hands awhile. "But it's one thing to come up with theories, and another to figure out what to do at breakfast," he said. "Someone had to meet the demands of research, and someone had to meet Genie's therapeutic needs, and I had both roles. And I was always aware that it was tricky mixing the two. I had a lot of ambivalence about it, at times. But in terms of the way we treated Genie—the things we did—I think we did about as good a job as anyone could have done.

"As far as the complexities of the case went, I wish they hadn't been there. In my hopes, I was blind to the complexities. They inhibited me from working right. There was no way of getting informed consent here, which has become a byword in human research. Genie never gave any indication that the filming or other activities were an imposition. If she had, we would have cut them out. Occasionally, we would get signs that she was stressed by the testing. But it's just like children's anxiety when they go to school for the first time: when they come home, they're very proud of themselves. Genie had a sense of triumph at doing many things for the first time. People don't grow when they're wrapped in cotton wool. They grow when they confront the world.

"The negative interpretations of the case are oversimplified, from my point of view. My own position—if I can psychoanalyze myself—was not one of expectation but of hope. The sky was not high enough for my hopes, but my expectations were down to earth. One easy out would have been for me to say early on that I would be much less involved. If I'd known what the outcome would be, I wouldn't have touched it—the outcome in general, and for me."

Other members of the Genie Team feel as bruised as Rigler does. They have imposed what amounts to a gag order on themselves and speak of the case reluctantly. As a result, a prominent piece of science has been forced into the shadows. The scientists have, by and large, lost contact with each other. Well before Irene's lawsuit, the grand collaboration established that early hopeful winter in Childrens Hospital had collapsed. The mixture of ambition, charity, and inquisitiveness that Genie attracted around her turned out to be explosive.

"Everybody got to be everybody's antagonist," Jay Shurley told me. "So many charges and countercharges. That's not what happens in a well-run scientific enterprise. Usually researchers grow closer through their association. This was a case of centrifugal force."

Some researchers, of course, were never very central to the research, and some, like Shurley, were flung to the periphery early on. David Elkind was one of those who might have been at the heart of things. He had known the Riglers from the sixties, when he was running the Child Development Center at the University of Denver and Marilyn was working with the hospital's nursery school. The Riglers invited Elkind into the May 1971 conference; afterward, he opted for an early departure. "I said, I don't want anything to do with this," he told me. "It was a battle. The child was the center of people's careers and grant-getting. I had trouble dealing with the clinical tests they were proposing. I felt so strongly that this kid just needed time to explore, but there were too many egos involved. The child was getting lost. I felt very bad about that. I got out."

Other scientists who had met Genie in those first glad months were upset by rumors of her later misfortunes; often, the rumors originated with Jean Ruch. "I was grateful to receive the various documents and accounts on Genie's sad life and the incredible behavior by members of the 'helping professions,'" read a reply letter Ruch received in 1984. The scientist who sent it had been a colleague of Eric Lenneberg and had been listed by Rigler as a consultant in the original NIMH grant proposal. "An exposure of the venal and fraudulent characters that are picking over the bones of this poor woman for their self-aggrandizement is certainly important from an ethical perspective, although I doubt that anything you might say will stop them," he said, and encouraged Ruch to write an account debunking the "nonsense" the scientists had been publishing. "Perhaps in this way, the lies will not be perpetuated from one textbook to the next, but I am skeptical."

The purported lies were largely, of course, those revelations published in Curtiss's dissertation. If Rigler's perceived failures were enduring targets for Jean Ruch's disapproval, Curtiss's success was no less so. Even after the bureaucratic acrimony had permanently separated Curtiss and Genie, Ruch hounded Curtiss, calling her at home and attending her professional lectures to pose hostile questions. The Wild Child thesis had been cleared in court and honored in numerous classrooms, but it and the linguist's further papers were condemned in Ruch's correspondence. Somewhere in the collective scientific mind, the charges stuck, and there are those in academe, still, whose acceptance of Curtiss's research is sullied with Ruch's contempt.

After her husband's death, Jean Butler Ruch continued to live in a beach house they had bought in Santa Monica. On visits there with her mother, Genie would stand inside the sliding glass doors, her hands held up before her in her persisting bunny posture, and watch the waves that had once so frightened and delighted her.

Ruch's letter-writing continued; the campaign culminated in her plans to write a book with Jay Shurley, setting the record straight. "I was bent on revelation," Shurley says. "She was bent on revenge." Their collaboration produced one paper, which Shurley and Ruch presented jointly at a Stanford University conference in 1985. There, Shurley was so upset at his partner's personal attacks on Genie's former caretakers that he resolved to distance himself from her further efforts. His desertion earned him status as the last entry on Ruch's long list of enemies.

Ruch's protests were cut short in 1986 by a stroke, the result of the vasculitis she had suffered since her school years. It left her aphasic, unable to coherently speak. Believers in a just fate or a personal God might have found her final torment too appropriate to ignore—the strident voice made mute, the compassionate urge crippled by the very childhood disease from which the compassion had been learned. A further stroke killed her in 1988.

37

Contrary to Ruch's assertions, Curtiss's research on Genie has proved its utility. It has been a platform for further observation, which in turn has given credence to Genie's revelations. "[It] was one of the first times scientists had used a case of an atypical child to understand the typical," Curtiss told me. "During the Genie research, a lot of other projects of that sort started."

One project that could be considered a direct descendant of Genie centered on a little girl named Marta. She was studied by Jeni Yamada, a linguist who as a graduate student helped Curtiss gather her data on Genie and helped Curtiss lobby for

Genie's welfare during the late 1970s. Marta was raised by caring parents but was victimized by a strange mental condition. She was Genie's opposite in that her cognitive abilities were crippled but her language was not. Her IQ stood in the forties but her syntax was respectable. She did what Genie could not—phrase impeccable sentences. But she had no semantic ability and could not, as Genie did irrepressibly, express an idea.

"Her talk sounded like mad-libs," Yamada told me. "Syntactically her utterances were fine, but they had no content. She would construct meaningless sentences perfectly. Some of them were too bizarre, but they were intact." Marta—in her late teens when Yamada knew her—would become transfixed with a color name or a number or even a part of speech, which word she would use over and over through a day's worth of sentences with no idea what real color or actual quantity it referred to. If the object of her fascination was the word "after," she might rattle on about "My uncle who used aftershave died after a heart attack after tennis. Afterwards . . . ," though her uncle was still alive and had never played a sport. She described emotions she did not feel. "She would say she was sad when her parents were gone or when John Lennon died, but it wasn't convincing. Her parents would come home and she would ignore them," Yamada said. "I think she enjoyed our relationship, but you couldn't tell it to look at her. She was better at talking about her emotions than feeling them. Genie, by contrast, had a real presence about her. She was endearing. You knew she was happy to see you. You could tell that she and Susan really had come to love each other."

Marta's contrasting handicap helped confirm what Genie's

handicap implied—the separation of cognition from language. "Theories that account for language by nonlinguistic abilities such as social interaction and so on are refuted by Marta," Yamada told me. "Those things—social interactions, cognitive ability—definitely enter in, but they don't account for language. There is something special in the language faculty, something unique."

Curtiss is still stalking that unique quantity. Genie's revealing profile—being so clearly able to think, so brilliant at nonverbal communication, and yet so utterly unable to master the rudiments of language—sent Curtiss questing for people who had other "selective deficits," or, like Marta, "islands of ability" that might shine new light on grammar's independence in the mind. "With Genie I had one facet of this relationship between language and cognition," Curtiss told me. "I wanted to find cases with a flip side of Genie's deficits."

To that end, Curtiss studied the demented elderly, individuals who are losing to senility their ability to communicate. "Their vocabulary becomes more and more inexact," Curtiss told me. "They talk about 'stuff' and 'things.' Their speech loses its melodic aspect. But their syntactic structures are intact."

Curtiss has also worked with retarded children. In recent years, she has consulted on the case of a woman who appeared retarded: Chelsea, a classic example of deafness mistaken for mental deficiency. "She's a better case than Genie," Harvard's Catherine Snow told me. "Chelsea is neurologically intact. She's deaf and grew up in the backwoods. So even though she had a caring family, she was deprived of language until she was in her thirties." At that age, finally, Chelsea was correctly diagnosed, by neurologist Peter Glusker, who has directed her treatment since and once a year brings her to Salk and to UCLA for testing by Neville and Curtiss. Even without the contamination of emotional trauma, Chelsea's brain response echoes Genie's. Her left hemisphere shows no specialization; she is semantically fine but cannot conquer syntax.

The corroboration does not surprise Curtiss. "Language is not a question of emotion or motivation," she told me. "It grows. Because it is the way it is, the question of how language is acquired is a lot more interesting. Language grows like an organ. When it comes to physical growth, no one asks why— why do arms grow? Learning a language is like learning to walk, a biological imperative timed to a certain point in development. It's not an emotional process."

Curtiss's most recent subjects are hemispherectomied children, who have had diseased or damaged halves of their brains surgically removed. The disability Genie suffered functionally, these children suffer literally. "I'm interested if hemispherectomied children can learn grammar," Curtiss told me. "Is the left hemisphere essential to fully realize grammar? The question is, when the left hemisphere is removed, how well can the right hemisphere do in supporting grammar functions?"

To find out, she evaluates children on the eve of their surgeries, then follows their recoveries. In surgeon's parlance, the operations are known as "morbid events." Curtiss's introductory interviews are called "premorbid documentations." I recalled that this was a woman who had said of her younger self that hospitals were not her strong point.

One summer evening Curtiss and I sat at a table in the kitchen of her house, talking about her recent work, while her husband, John, readied dinner. The house was a modest clapboard bungalow a few blocks from the Santa Monica Freeway, in the vast Los Angeles flatland. The soupçon of yard outside would not have accommodated a volleyball game. At the table, Curtiss spoke of S.M., a hemispherectomied boy we had visited earlier in the day in a poor suburban neighborhood not far from Genie's Temple City home.

"His profile doesn't fit what we thought," Curtiss said. "He understands perfectly fine. He knows his grammar. S.M. suggests that right hemisphere potential for mastering grammar is pretty complete, if used early enough.

"This isn't all that surprising," she continued. "Language is one of the key abilities that allows us to succeed as a species. It stands to reason that the brain would be malleable enough to protect this ability, however it can." She credited S.M.'s adaptability to the influence of an outside agent more personal than brute biology. "His mother is an inspiration," Curtiss said. "We always leave their home awed at her. She has no resources, no money. But she'll take on the world. She's a fighter. She's responsible for a lot of his progress. She's driven him to overcome his problems." S.M.'s mother has even pushed him to overcome the physical paralysis that usually ensues from the surgery, even though the motor system is even more intractable than the language and cognitive systems of the brain. "It's *impervious* to change," Curtiss said. "But S.M. has no paralysis. He's overcome."

Wasn't it remarkable that such a recovery could be the product of hard work? I asked Curtiss. Didn't it seem that the brain would repair itself or not but wouldn't be amenable to coercion? Even S.M.'s language learning, a process not controlled by emotion, seemed marvelously susceptible to the influence of a nurturing hand.

Curtiss jumped from her seat to intercept her four-year-old, who was hurrying to assist her father as he skinned and boned a chicken at the sink. She sat back down with her daughter squiggling captive on her knee and her answer in mind. "It's unusual," she said. "That's true. But there's a whole world of unusual things that people will tell you about that are beyond medical explanation."

38

"Beyond medical explanation," as it turns out, is a notion I had heard before, from another scientist, in a not so dissimilar room, on the day I took David Rigler's advice and went to see Jay Shurley.

Shurley's study is an aluminum-sided sun porch tacked on to the back of his home in Oklahoma City. The place seems unremarkable until you have been there awhile, and then you notice that it is as populated as an ice floe with penguins—tiny ceramic ones march across the window ledges and are posted like sentries on his desk; there was a penguin logo on Shurley's golf shirt and, when he served me coffee, there were penguins printed on the cup. On the wall is a map of Antarctica next to a map of Texas—the geographies of his two great encounters with isolation, the one he studied as a scientist and the one he endured as a child. Through the open doorway leading to the backyard I could hear the tinkling of wind chimes and the constant chirping of finches in the silver maples.

When I had accepted the coffee, Shurley left the room and returned with two great cartons marked PEPSI-COLA and filled with manila folders. Displacing a penguin, he laid them on the desk. They were his Genie files. As he talked with me during the next several days—beginning after breakfast, ignoring lunch, and continuing on through dinner—he would dip into the boxes for letters, symposium papers, the scribbled logs of phone conversations he had had with Rigler, Ruch, Kent, and Hansen almost twenty years earlier. There was a file marked "Sleep Spindles" in one of the boxes, but by and large what he had preserved in his cardboard repository was not the science of Genie but the experience. The question that tormented him lay somewhere beyond the data.

"Here," Shurley said, reaching into a carton. The files were labeled "Genie Emerges," "Jean's Input," and "Genie Book" (in the outline of which Genie's life was divided into "Genesis" and "Exodus"). He pulled out a file labeled "Photos."

The first picture he handed me was of a nondescript house, seen from across a street through a picket of royal palms. Pages of a newspaper blow across a yard through the cold gray shade of a lemon tree. A second photograph was of the same house, but it was taken from the drive, where Irene stands in a plaid skirt and holds a cloth purse tight against her smooth yellow cardigan, as though expecting a sudden chill. It was the December day, soon after her acquittal on charges of child abuse and neglect, when the house was first opened for inspection by curious strangers.

"Irene had all the instincts of motherhood, to my mind," Shurley said. "And she was very thwarted, and she was very weak. Only after a long period of befriending by Jean Ruch was Irene able to stand up and reassert herself. I remember some years ago, when she was living in almost abject poverty, one of the big networks—maybe overseas—came along and offered her $10,000 for the story, and put all these documents in front of her, and she told them firmly, 'No.' I was there at the time—at least, I was in Los Angeles and talking with her— and I was amazed at the strength of her fear, or the strength of her conviction."

Shurley set the pictures of the house aside and drew a rectangle on a piece of notebook paper. He divided it up into smaller rectangles. "Here is the room they said was a shrine to Clark's mother," he said. "It was the master bedroom, and it was almost completely filled with the bed. It wasn't very large. Here's the living room, and there was a chair here, and the television, which didn't work. Clark slept in the chair most of the time. He slept there, and here is the pallet where his son slept, on the floor." He drew a square in the corner for Genie's room. "She had a window here, and another around the corner, over here. The dresser was here, between them, and here is where she slept." He drew a small rectangle and labeled it CRIB. "And here is the potty chair," he said. "Sometimes it was over here." Shurley looked up and then back, and drew a yard around Genie's house, with a driveway and a lemon tree. The lemon tree, from above, was a cloud with dots.

The next several photographs were taken on that same winter day, but they were taken inside, in Genie's room. The room was dim. Here were the closet doors—three plywood panels with chrome pull handles. The dresser was pine and had four drawers. And here were the two windows, the upper half of each covered by a shade. Yellowish half-curtains draped the lower halves, their fabric thin and patterned with red flowers. One window's curtain had been pulled back and was fastened to the wall with packaging tape.

"Genie's room was not sensory deprivation so much as sensory monotony," Shurley said. "Monotony. You know, variety is not the spice of life; it's the very stuff of life. To the development of a defensible, adaptable ego, monotony is deadly. In that little room, a person would project internal images, not absorb outside ones, and would become confused about what was real and what was imagined—would lose the ability to differentiate between dream and waking. Socially isolated children usually have psychotic parents who treat them as animals. There is no encouragement of any human closeness. It is typical for them to be locked in a closet—it isn't rare. There was a boy here in Oklahoma City recently who was four years old, and his parents were keeping him penned with the dogs in back of the house. He walked on all fours. Genie remains by a good bit the kingpin of these cases. She has the record. Though it's not a record that anyone would envy."

The next photograph had been taken half a year later. It was summer, and Genie was sitting on a floor, laughing and alert. A note on the back of the photograph read, "This photo was taken about three days after she came to stay with me (she has hospital p.j.s on)." The note is in Jean Ruch's hand. "The ability of that little girl to elicit emotion on the part of the observer was fantastic," Shurley said. "You had to witness it.

Just hearing about it would be orders of magnitude from the actual experience.

"Jean and Floyd Ruch, they were almost obsessed with this child. Jean really did latch on to Genie in the early days, and it was reciprocated. Jean, of course, had never had a child of her own. Rigler had three and felt that experience was on his side. But after I got to know Jean I didn't see anything to suggest that she wouldn't be a good foster parent. She was the teacher, and had developed a very positive relationship with Genie within a couple of weeks. I never found the Riglers to be that warm or empathetic with her. At their house, it was as though Genie were being studied in a cold frame rather than in a hothouse. I understand some of Rigler's feelings about Jean Ruch. She had a very interesting paradoxical streak: she could be extraordinarily kind and sensitive to children—and she was, as teacher to some very disabled and sick children—and then she was capable of doing malicious and, I'll say, sadistic things, not to the children but to those who she felt were in disagreement with her about how the children should be treated.

"But to several of us, it seemed a pity that Genie could not be with someone like Ruch, who would bond to her as a person and not as a scientific case. Besides, I tend to go with the child. If the child says, 'I like this person,' there's something real there that a child can latch on to. To adults there may be things that don't seem right, that cause concern. But the child's instinct is usually right on the issue that's most important."

There were a few other photographs from the summer of 1971: Genie at an art gallery, stepping into a patch of bright sun in a smart maroon dress with a white collar and big white pockets; Genie in a swimsuit at the beach, concentrating with apparent delight as a receding wave washes around her feet, and holding her hand up in the OK sign, the tip of the forefinger joined to the tip of the thumb.

The last two photographs were of someone else, or so I thought: a large, bumbling woman with a facial expression of cowlike incomprehension. In one picture, the woman sits in a car pretending to drive, her eyes at half mast, her front teeth protruding in a drawn grin, a starburst reflection of palm tops floating in the windshield glass. In the second, the woman is indoors. She is about to cut a birthday cake with white frosting. Her eyes focus poorly on the cake. Her dark hair has been hacked off raggedly at the top of her forehead, giving her the aspect of an asylum inmate. Something about her dress is sad and reminiscent: it is shapeless and has red flowers. Her right hand grips the cake knife, and her left hand is held in front of her, forefinger touching thumb.

Shurley watched grimly as my recognition dawned. "Her twenty-seventh birthday party," he said. "I was there, and then I saw her again when she was twenty-nine, and she still looked miserable. She looked to me like a chronically institutionalized person. It was heartrending." A note by Shurley on the back of the photograph read, "Genie is very stooped and rarely makes eye contact. This photo was at her happiest, other than when momentarily greeting her mother and me an hour earlier."

As I turned the photograph back over, my association with the dress came clear to me. "Irene sewed it," Shurley told me. "She'd been a master seamstress before her eyesight went." The dress, its thin weight and floral pattern, reminded me of the curtains in the little room.

"What do you make of her expression?" I asked Shurley.

"What do I make of it?" he said. "She looks demented." He paused, and then spoke intensely, as though he were at the center of something. "The way I think of Genie, she was this isolated person, incarcerated for all those years, and then she emerged and lived in a more reasonable world for a while, and responded to this world, and then the door was shut and she withdrew again and her soul was sick." Without looking away from my face, he pointed to the photograph of the woman in the car. "This is soul sickness," he said. "There is no medical explanation for her decline into what appears to be organic, biological dementia."

39

For a while, Shurley seemed disinclined to speak, and we listened to the finches in the yard. Then he said, "At the time that Genie came to light, I went back to try to find, anywhere I could, any kind of directions. Anything that said, 'In case of tornado do this, in case of earthquake do this, and in case of an experiment in nature do this.' I found it nowhere. There's nothing of the sort. But from my experience the research with Genie could not have been handled worse. The process went off track from the day it was conceived. It went, after a little while, a hundred and eighty degrees from the direction it ought to have taken. There is a fundamental issue here that nobody has grasped. The key issue—I believe now, very strongly, in terms of my own experiences with isolation in many different contexts—is not the acute effects of the isolation. It is the problem of reentry into the matrix from which the child has been isolated. Isolation places one's own readiness to react in a kind of cold storage. Imagine using a muscle that has been in a cast, or a sling. Once you take the encumbrance off, the muscle has to retrain itself. It's suffering from atrophy, from disuse. Rehabilitation involves figuring out how you allow the strength back without rupturing anything.

"We're born helpless. We are born into the world with no boundary between self and not-self. We spend the first twenty years of our life establishing that boundary. Children who are so abused, deprived, are losing that battle by the age of three or four. I felt that Genie was one of those—a little girl with no sense of herself as a separate, inviolable entity. I wanted Genie to come into the world as a core ego, capable of trust and mistrust. Proper reentry is a key ingredient in treatment and in research. A proper reentry is not one preempted by scientific exploitation gone wild.

"A child needs more than approval. She needs a sense of security, safety—the absolute conviction that she is worthwhile. Well, Genie grew up in a house where the father didn't like himself and the mother didn't like herself and no one liked Genie. And later she was a celebrity. All these people looking at this extremely primitive child—this larval child. In this six-year-old body, a thirteen-year-old girl. Talk about a weird kid: Genie was a weird kid. And that's how she was treated by everyone—as a weird kid: 'What do you do with a poor, weird kid like that?' Genie was viewed as a child views feces—first as treasure, then as shit, in Anglo-Saxon terms. And, really, what did Genie, taken apart, have to offer the world? Except for her unique early-life development, not much. Not much.

"Genie's problem was seen too much as a pedagogical one, not an emotional one. We tried to teach her language. Well, I don't know. There's a problem. In Linnaeus's classification, *Homo sapiens* is known as *cultura,* not as *lingua.* Our advancements take place in a relationship. In order for an infant to learn anything—and this takes you back to Victor, the Wild Boy of Aveyron—there has to be a relationship in which the child gets enough nurturance to proceed. Affective attachment plays the primary role. It is not an intellectual process. Intellect rides on the back of affective bonding. And affection's not easy to come by. Human beings have a unique talent not only for cruelty but for indifference. Compassion was not referred to by the Enlightenment philosophers as the essential or defining characteristic of humankind. It's something in our nature that must be taught."

Shurley waved a hand dismissively. "This is old stuff," he said. "I resolved that if I lived long enough I would do a case study that would show how things should be approached in cases like this. These experiments come along. Victor in 1800. Kaspar Hauser in the 1820s, I believe. Genie in 1970. None of the wild children have been handled well. All of them were handled the way Genie was. She *could* have been handled well. She would have been a disappointment in some ways, but the outcome would have been happier, certainly. Genie arrived at the hospital, and within the first couple of months she became hungry. She came out of an environment that was unfriendly but consistent. Now she was in a new environment, with noise and other kids. A hospital is an overstimulating place. The problem was how to get her out of it and into a home. But she went from one home to another. More noise. She went from famine to feast. Her response was not to take that feast. She was overwhelmed. This is part of the emergence thing. She was enormously starved, but the starvation was so chronic, so long-lasting, that she didn't trust her world to give her what she wanted. She was afraid that part of what she would be given would be toxic to her. As it turns out, she was right.

"These were not bad people. They just didn't allow this child to develop along normal lines. The course of research defeated the treatment, which defeated the research. The science would have fared better if the human aspect had been put first. We probably would have learned a lot more, and what we learned would have been transferable to other cases. The only generalization you can get from this is as a bad example—an example of how not to do it.

"What I saw happen with Genie was a pretty crass form of exploitation. I had to realize that I was a part of it, and swear to refrain. It turned out that Genie, who had been so terribly abused, was exploited all over again. She was exploited extrafamilially just as she was exploited intrafamilially—just by a different cast of characters, of which I'm sorry to say I was one. As far as Genie is concerned, it's a fated case. You have a second chance in a situation like that—a chance to rescue the child. But you don't get a third chance, and that's the situation now. We can't do the experiment over. We can't go back. And that's the bitterness."

40

The story of Genie, as I received it from Jay Shurley during those days in Oklahoma, and received it from others over the course of several years, has come to seem to me a history of transcendence. More accurately, it is twin histories: that of a young woman trying desperately, heroically, and ultimately unsuccessfully, to transcend the confining horrors of her childhood, and that of researchers intent on transcending the limits of what they knew, within the confining integrities of their science. What the scientists wished to discover, Genie wished to claim—a human birthright, a key to our essential, abiding natures. In that mutual quest, the successes and failures of scientist and subject became linked, linked as inextricably as the blame and virtue, hope and futility that still attend the case in the minds of those who lived it. I knew, those days in Oklahoma City, and would be reminded often again, that Jay Shurley's view was not everyone's view—that there were those who had found the Rigler home warmer or the science less obtrusive, or who saw Jean Ruch or the social workers as more single-handedly the shapers of Genie's decline. In the diaspora of opinion surrounding the case, I would not find one among Genie's observers who would agree entirely with any one of the others. There were some whose only quarrel with Shurley's critique was with its moderation. Among the many I talked with, the only great consensus was one in which Shurley shared. The expression that crossed his eyes, as he spoke, of anger and imploring was no different than the one I had seen in Rigler's eyes, the look I would see in Curtiss's, not for the first time, during our last evening interview in her home.

It seemed to me tangible that for those whose careers had been made by Genie and those whose careers had been broken, the time with the child occupied a place outside career, an event in their lives as central and defining and large beyond appraisal as first loves and the births of their own children. Genie had appeared to them out of the little room, and they had thought in the manner of scientists that here was someone of whom they might ask questions and who might be able to respond. They thought, in the manner of scientists, that these were questions that could be asked impartially, from without, and without emotional risk. From their safe remove, they set out to explore this other species, this Irene and Clark and Genie and their sad, unimaginable history. Later, though, the histories became mixed, those of the observers and those of the observed, and no longer was there any safe remove. It turned out the scientists had not freed Genie from the little room; instead she had ushered them in. Ushered them in and abandoned them. Genie had passed through their lives like some excoriating prophet and left them with an icon on an office divider and a cardboard box of relics, and a question of isolation and captivity that would not have been nearly so bedeviling if it had not been asked about themselves.

Around the table in Curtiss's home, dinner was done. John had lured their two young daughters away so that Curtiss and I could talk. The drone of a television summer rerun and an occasional fit of giggles escaped from an unseen room. Watching Curtiss with her daughters and, earlier in the day, with her subject S.M., I noticed that children drew an easy, playful kindness from her. Curtiss is, in any case, a person of unsuspected softnesses. She had told me firmly when we first met that she would talk only about science—that her personal history with Genie was out of bounds. But at the end of that interview, and of each thereafter, she violated her own restriction and, without prompting, spoke movingly of her feelings for the child she had investigated. "I developed a need for her," Curtiss would say. "I missed her when she wasn't in my life."

Over our meal and dessert, and now over uncleared dishes, Curtiss and I concluded our final hour of syntax and semantics, critical periods and hemispherectomies. As I closed up a notebook and put away a pencil, she veered again out of the confident realm of research and into that forbidden personal room. There was desolation in her voice. "I would pay a lot of money to see her," she said. "I would do a lot. I haven't heard from her in years. And I've heard only two reports. The last one was that she was speaking very little, that she was withdrawn, depressed. Genie was very lovable. She was beautiful.

When John and I first met, I would tell him about her, and he would say, 'Stop! Stop! You're building this person up so much that if I meet her I'll be disillusioned. No one can be that wonderful.' Then he met her, and when we left he said, 'My God! Why didn't you tell me?'"

Curtiss's older daughter pirouetted into the kitchen to show us her sunglasses. The earpieces were gone, and the lenses, perched on her nose, were valentine-shaped. She leaned into her mother, and Curtiss put an arm around her shoulder. But Curtiss's mind was elsewhere, and the little girl skittered off, back into the recesses of the house.

"What is it that language can do for a person?" Curtiss asked. "It allows us to cognize, to think, and that's important to me, because I'm that type of person. It also allows us to share ourselves with others—our ideas and thoughts. And that provides a huge part of what I consider to be human in my existence. Genie learned how to encode concepts through words. She used language as a tool: she could label things, ideas, emotions. It afforded her a completely new way to interact with her world. If I had to choose the pieces of language that would serve me best in being human, they would be the parts Genie had. It was from her we learned of her past. She told us of her feelings. She shared her heart and mind. From that perspective, who cares about grammar? Acquiring those parts of language didn't cure her. She's unbearably disturbed. But it allowed her to share herself with others. For years after I was not permitted to see her again, I would wonder about what I would say to her if I saw her. Not just how I would react—I know I would give her a hug—but what would I say?

"Genie is the most powerful, most inspiring person I've ever met. I'd give up my job, I'd change careers, to see her again. I worked with her, and I knew her as a friend. And, of the two, the important thing was getting to know her. I would give up the rest to know her again."

AFTERWORD

Every book contains two stories. The first is the one it purports to tell and is necessarily noisy—trumpeted by the author on every page, and also on dust jacket and press release by publisher and publicity agent. The second is the story of its telling, and it is generally more quiet, sometimes going unremarked even by the author, of whose experience it is formed. There is a relationship between the two, as intangible and inescapable as that between those twin stars of which astronomers tell us, where the dark star is visible only through the wobbles it creates in the orbits of its brighter, larger twin. The strength of this gravitational pull—the depth and vitality, urgency and honesty of the connection between the two stories—goes a long way toward determining how close a book will come to being what we call literature.

I was surprised, after *Genie*'s first publication in the United States, at the number of questions I received, not about the brighter, larger story, but about the hidden one, about my own personal experience. How had I learned of Genie? How had my immersion in her life changed mine? And, most insistently: Had I met her? And how had she seemed? My responding (unspoken) question was, How had my readers known to ask about this? For indeed Genie had affected me greatly, in ways that, by changing me, changed my book as well. I chalked up their awareness as one more surprise for which my inexperience in book writing—for *Genie* was my first book—had left me unprepared.

It had left me unprepared for plenty else, and especially for the way a book, like a child, deranges the life of its parent while willfully assuming a wayward life of its own. Like a child, the story of Genie escalated slowly into its final full stature. When I first heard of the case, from a linguist at the University of Illinois, it was only in a passing sentence, and had a cast no larger than that of a Christmas crèche in a suburban schoolyard: there was a mother, a father, a child, and an indistinct halo of wise men and animals, easily confounded. My early interviews with Genie's scientific magi gave me a glimpse of the guilt, anger, and fear lying beneath the surface sheen of professional composure, and I began suspect the story's depth and size. I proposed it to *The New Yorker,* and promised its editors 20,000 words in nine months. Three years later, I gave them 70,000, and knew I would expand that into a book.

The transformation of the story during those years is evident in the consequences of two phone calls I made, one near the beginning of my research and one at the end, and both to Irene, Genie's mother. Talking with Irene, and obtaining her permission to talk with Genie, had been a tantalizing and much dreaded prospect since the beginning of my research. It had not been easy even to find Irene; she had done her best to disappear, hiding her whereabouts from everyone who had known and worked with her daughter. In tracking her down, I was compelled by more than curiosity; the future of my project, it seemed to me, was at stake. That's because my project was in trouble. None of the scientists on whom I relied for my portrait of Genie had welcomed my intrusions; for them, my questions tore the scab off a very painful wound. Some refused to talk, and those others more generous gave me accounts that were without exception at odds with the accounts of their compatriots, for the same anger and fear that closed them off from me had separated them from one another. It was a journalist's nightmare, a host of competing and conflicting realities, all strongly believed in. Thinking that the only person able to settle this clamor was the one in the center of the story, I made my first call to Irene.

It went disastrously. The woman who answered the phone was so adamant about not talking with me that she talked very little even in turning me down. "Not interested," Irene said, and repeated the phrase again with each of my stumbling attempts to make my proposal sound attractive. Not interested in seeing me. Not interested in letting me see Genie. Just plain "Not interested." As I hung up I was swept with discouragement, a mood that dogged me still, four months later, when I gave up my labors for lost and moved from California.

My move was a flight, simple enough, and I was fleeing more than the wreckage of my story. During those several years, my life had changed in ways that would rate as banal in any compendium of writerly angst—a relationship ended, a bank account emptied and, finally, an attempt to escape it all and run. I can't say why I chose Paris as my hideaway. The city had a slim connection to Genie's story in that her important precursor, Victor, the Wild Boy of Aveyron, had lived there in the Institut National des Sourds-Muets during the first years of the nineteenth century. But Victor was far from my mind.

Perhaps it was just that Paris had had a hold on my imagination and affections ever since my first acquaintance with it, years earlier. I had, however, never spent a winter's day in France, and winter was full upon the capital when I arrived. It was high season. A cold damp melancholy had settled over marble façades and stone streets, and if the city's heart was lit with a warm interior intimacy, it was an intimacy I viewed from without, yellow lamplight through window glass. I found my first benefits in that melancholy—it was as receptive of my dark mood as the sun of California had been mocking, and so a solace. For my first weeks, I lived in an apartment lent me by generous and absent friends near Buttes Chaumont, a grand and ascetic apartment, cavernous and empty, with only a bed, a chair, and a table for furniture, and an extaordinary view over a city of five million strangers. In a bookstore, I lucked into an acquaintanceship with one of those strangers, who kindly arranged for the home I rented for the remainder of my stay, a tiny *chambre de bonne* on rue Henri Barbusse, near the Observatoire, in the Fifth Arrondissement.

I thought it lucky that my new home should be on a street named for a journalist, and a beleaguered one at that. But I could not have foreseen the true magnitude of my good fortune. The particular *quartier* was not one I had ever explored—I arrived in it at night, following scribbled directions to the Port Royal R.E.R. stop and down through dim steets past shuttered *boulangeries* and *charcuteries* and a closed *lycée,* carrying my suitcase of clothes, and my other, heavier suitcase filled with *Genie* notes. I had lugged those notes from the United States in an effort to uphold at least the outward appearance of being an employed writer and to defend my private pretense that I was not running away. Since my arrival, the bag had not been opened—truly I had no idea what I might do with those notes; they seemed so distant from me, more foreign to my experience and endeavor than anything else in the foreign city around me. At best they were a burden. That I did not drop them over the Pont Neuf showed more respect for the Seine than for the story I had given up telling. But there was another aspect my inexperience with book writing had not prepared me for, and that was how luck shows up like a benevolent stranger to lead a lost story back home.

The next morning I took my first tour of my new residence and the cobblestoned *cour,* with its flanking of apartment doorways and the glass facade of a gilder's atelier. There was an elm tree in the *cour,* and its yellow leaves on the paving stones had been swept aside by passing feet to leave a dark path toward the door to the street. When I pushed open that heavy green door, and stepped through onto the sidewalk, I received the shock that changed my experience in Paris and the future of my book. Facing me across the narrow prospect was a broad garden and the white, many-windowed façade of a building that I did not know even continued to exist, though I had seen it in enough old engravings to recognize it anywhere: the Institute National des Jeunes Sourds (as it is now called,) the home of Genie's forerunner, Victor of Aveyron. I had somehow contrived to become Victor's neighbor, a serendipity I took as a sign that, whatever my intentions in coming to Paris, I had not been meant to escape, that my story was here, and that, however inadvertently, I had found it. Or perhaps it had found me. Either way: In that moment on rue Henri Barbusse, I knew the book would get done.

Over the next months, I began the serious construction of that story, working away on a laptop computer (ever the conspicuous American) in the Bibliothèque Ste. Geneviève, Centre Pompidou, and the beautiful high sanctuary of the reading room of the Bibliothèque Nationale. The advantages of the city to an American writer were manifest: I knew no one, and my French was not good enough to allow me to eavesdrop, so I could carry a single thought in my head for an entire day, through any sort of crowd, without distraction. My *chambre de bonne* was too small to pace in, so I sat, my writing spread on a folding metal tea table: on the occasional banner day there I slaved out 5000 words before walking out at midnight to reward myself with *fruits de mer* or *steak tartare* at the *brasserie* of the Closerie de Lilas.

Later, when some months and winter had passed and the time had come for my return to San Francisco, I knew that I had broken the story's back, as they say. Sitting in my little room, alone in a world whose language I did not speak, I had watched my story attain a certain sure pacing and confident tone. It had developed something more, too, though only in retrospect could I see what that quality was. In my fevered isolation, I had come to understand my story in a new way, had for the first time come to grips with the part of it that had gripped me from the start. I think it must be true of all writers that while involved in a passionate project they lead dual lives, the one in their head and the one in public, each populated with its own people and events, emotions and peculiar demands. Inevitably, the two worlds mingle, and each life is lived partially in the other's realm. So it was with me, and in the mingling of my daily life and Genie's history, I had developed an understanding of the story that had earlier eluded me. A half a year after I returned to the United States, I had a manuscript in hand, and, more importantly, in the mail to *The New Yorker*, and it was then that I made my second call to Irene.

My intent was simply to tell Irene that the story would soon be published. I thought I owed her that. I fretted over the call, and paced and practiced before picking up the phone because I wanted to be sure to get my message across before she cut me off, as I knew she would. But she didn't. Far from it. Instead, she received me graciously and talked with me willingly for an hour and a half. To complete my astonishment, she invited me to her home for a visit. Visit I did, again and again, and the conversations between us over the course of the following summer, along with the many documents she allowed me to read, provided much of the material that inflated the original *New Yorker* pieces into this book. Talking with Irene was, withall, an odd and overpowering experience: Here I was, in the presence of a person whose life had consumed me for years, but whom I had not expected to meet. If I had been writing a biography of Martha Washington, I could not have been more delighted and horrified by the sudden luxury of meeting my subject in the flesh.

We settled quickly into a sort of easy intimacy, I and this tiny woman, stooped with age, her movements of hand and foot made exquisitely precise and vaguely questioning by a life of episodic blindness—for by now her sight had been robbed again, this time by glaucoma. I would buy Chinese takeout food on my way to her home, for which she would insist on paying me back from a plastic change purse, and we would sit afternoons at a kitchen table in her meticulously neat apartment in the dry desert heat of southern California, and talk as though the central facts of our meeting were not true, as though I were not really a reporter and she were not really a woman who had lived a dozen of her married years with her daughter tied prisoner in a closed back bedroom. The daughter she described to me is doing better now than at any time in a decade—she is vastly troubled and generally cheerful, and she talks, though not very comprehensibly. If she has nowhere reached the potential envisioned during the hopeful early days of her emergence, Genie has nevertheless improved from her most desperate state. She has entered a new and more supportive foster home, and she still visits Irene regularly. Irene's openness left me with a decision to make: Should I press her to let me meet Genie, to accompany them during one of those visits together? It was my old dream, encountered at last, and after some thought, I turned the possibility down.

I could not, after all, simply meet her as myself, as a friend; inevitably I would be an ambitious journalist, taking notes, alert for an answer. How could I not help but open another chapter in Genie's long history of exploitation? And what would I do with my answer, once I got it? There is an impatience in modern times, and especially in America, with the true stories of things: they are too ambiguous and untidy and inconclusive; they suggest more mysteries than they solve. We prefer to have an answer—a solution, thank you—no matter how implausibly simple. And so we like to see things directly, in the flesh, to gawk when we might listen. Indeed, the less comprehending of my reviewers would later call out for me to meet the child I talked about. But just as strongly as I had once wished to see Genie, I now saw the desire as prurient, and also self-defeating. If I met her, how would I be able to hold my experience with her back from the reader, without seeming intolerably coy? And with Genie's blinding presence front and center in the stage lights' glare, how could my reader then discern the shadows I was trying to explore? It turns out my story lay elsewhere than I originally thought; I now believed this tragic girl would best be perceived and portrayed through all the impressions and memories, all the hopes, fears, ambitions, courage, cowardice, anger, and love of those people who had known her during her emergence, and whose job it had been to bring her back into the world. In short, I wanted my reader see Genie not directly but in reflection, in both of the meanings of the word.

In deciding that, I turned much of my tale over to the scientists and caretakers who had been so opposed to my undertaking, and who would be, some of them, enraged at the results. The violent discrepancies of their accounts, those competing realities that had once caused me such despair, now seemed proof to me of Genie's ultimate significance. The question the scientists originally asked of her—"What does it mean to be human?"—is one that Genie eventually asked of them, and their answers were evident in the effect she had on their individual lives. How, then, could the answers not be as various as the lives she had entered and transformed? My understanding of this came naturally, for hadn't Genie undergone a similar transition within me, from being an object of scrutiny to being an inescapable presence whose story was intertwined with my own? So she had—during that winter when she and I paced the streets of the Fifth Arrondissement, that winter we spent interviewing in the Institut National des Jeunes Sourds and reading in the Bibliothèque Ste. Geneviève. That was where I had been given the key, the understanding that allowed me to tell her story. That was the revelation of Paris.

In the time that has passed since the first publication of Genie, I have been pleased at the changes the book has wrought. The scientists who had severed their association a decade or more ago are talking again; in some cases, they have hinted at renewed collaboration. There is also some hope—and certainly it is a hope of mine—that Genie may again emerge into the larger world, to her benefit. I helped to arrange a reconciliation of several scientists and Irene, and that is producing its own fruits, independent of any direct effort of mine: Last spring, Irene invited David and Marilyn Rigler to visit their onetime foster child. When Genie saw the Riglers for the first time in fifteen years, she called them both by name.

Over the months, I have received numerous letters attesting to what a writer hopes for most: that his story has reached someone. Readers write to ask me more about Genie, and some to ask if there's anything they can do. But many others write to tell me about themselves. This, too, has surprised me. A few relate sagas of child abuse or language handicap, but others with no personal history of neglect or novel deprivation nevertheless express an intense identification with Genie's story. I know what they are saying: that her life has entered theirs, that they recognize, in her singular and extreme misfortune, some universal condition. No one of us has suffered what Genie suffered, certainly. But who among us has not felt isolated in our past, imprisoned in our fate, helpless in finding the words to express what it is that has happened to us? Because of that, Genie calls each of us by name. That's what the letters tell me. In them, I have discovered the last thing for which my inexperience in book writing left me unprepared. A book tells many stories: That of its subject. That of its author. And, if it is to come anywhere close to meeting the demands of literature, those of its readers, as well.

Russ Rymer
Fernandina Beach, Florida
November 1993